Meaningful Stuff

Design Thinking, Design Theory

Ken Friedman and Erik Stolterman, editors

Design Things, A. Telier (Thomas Binder, Pelle Ehn, Giorgio De Michelis, Giulio Jacucci, Per Linde, and Ina Wagner), 2011

China's Design Revolution, Lorraine Justice, 2012

Adversarial Design, Carl DiSalvo, 2012

The Aesthetics of Imagination in Design, Mads Nygaard Folkmann, 2013

Linkography: Unfolding the Design Process, Gabriela Goldschmidt, 2014

Situated Design Methods, edited by Jesper Simonsen, Connie Svabo, Sara Malou Strandvad, Kristine Samson, Morten Hertzum, and Ole Erik Hansen, 2014

Taking [A]part: The Politics and Aesthetics of Participation in Experience-Centered Design, John McCarthy and Peter Wright, 2015

Design, When Everybody Designs: An Introduction to Design for Social Innovation, Ezio Manzini, 2015

Frame Innovation: Creating New Thinking by Design, Kees Dorst, 2015

Designing Publics, Christopher A. Le Dantec, 2016

Overcrowded: Designing Meaningful Products in a World Awash with Ideas, Roberto Verganti, 2016

FireSigns: A Semiotic Theory for Graphic Design, Steven Skaggs, 2017

Making Design Theory, Johan Redström, 2017

Critical Fabulations: Reworking the Methods and Margins of Design, Daniela Rosner, 2018

Designing with the Body: Somaesthetic Interaction Design, Kristina Höök, 2018

Discursive Design: Critical, Speculative, and Alternative Things, Bruce M. Tharp and Stephanie M. Tharp, 2018

Pretense Design: Surface over Substance, Per Mollerup, 2019

Being and the Screen: How the Digital Changes Perception, Stéphane Vial, 2019

How Artifacts Afford: The Power and Politics of Everyday Things, Jenny L. Davis, 2020

Meaningful Stuff: Design That Lasts, Jonathan Chapman, 2021

Meaningful Stuff

Design That Lasts

Jonathan Chapman

The MIT Press
Cambridge, Massachusetts
London, England

© 2021 Massachusetts Institute of Technology

All rights reserved. No part of this book may be reproduced in any form by any electronic or mechanical means (including photocopying, recording, or information storage and retrieval) without permission in writing from the publisher.

This book was set in ITC Stone Serif Std and ITC Stone Sans Std by New Best-set Typesetters Ltd. Printed and bound in the United States of America.

Library of Congress Cataloging-in-Publication Data

Names: Chapman, Jonathan, 1974- author.
Title: Meaningful stuff : design that lasts / Jonathan Chapman.
Description: Cambridge, Massachusetts : The MIT Press, [2021] | Series: Design thinking, design theory | Includes bibliographical references and index.
Identifiers: LCCN 2020025447 | ISBN 9780262045728 (hardcover)
Subjects: LCSH: Product design. | Material culture. | Commercial products—Psychological aspects. | Waste minimization.
Classification: LCC TS171.4 .C464 2021 | DDC 670—dc23
LC record available at https://lccn.loc.gov/2020025447

10 9 8 7 6 5 4 3 2 1

Contents

Series Foreword   vii

1   **Making and Breaking the World**   1
2   **Cultures of Keeping**   23
3   **Material Matters**   47
4   **Deeper Experiencing**   69
5   **Aging Spectacularly**   97
6   **Urban Mines**   123
7   **Design That Lasts**   149

Acknowledgments   167
Notes   169
Bibliography   193
Index   213

Series Foreword

As professions go, design is relatively young. The practice of design predates professions. In fact, the practice of design—making things to serve a useful goal, making tools—predates the human race. Making tools is one of the attributes that made us human in the first place.

Design, in the most generic sense of the word, began over 2.5 million years ago when *Homo habilis* manufactured the first tools. Human beings were designing well before we began to walk upright. Four hundred thousand years ago, we began to manufacture spears. By forty thousand years ago, we had moved up to specialized tools.

Urban design and architecture came along ten thousand years ago in Mesopotamia. Interior architecture and furniture design probably emerged with them. It was another five thousand years before graphic design and typography got their start in Sumeria with the development of cuneiform. After that, things picked up speed.

All goods and services are designed. The urge to design—to consider a situation, imagine a better situation, and act to create that improved situation—goes back to our prehuman ancestors. Making tools helped us to become what we are: design helped to make us human.

Today, the word *design* means many things. The common factor linking them is service, and designers are engaged in a service profession in which the results of their work meet human needs.

Design is first of all a process. The word *design* entered the English language in the 1500s as a verb, with the first written citation of the verb dated to the year 1548. *Merriam-Webster's Collegiate Dictionary* defines the verb *design* as "to conceive and plan out in the mind; to have as a specific purpose; to devise for a specific function or end." Related to these is the act of drawing, with an emphasis on the nature of the drawing as a plan or map, as well as "to draw plans for; to create, fashion, execute or construct according to plan."

Half a century later, the word began to be used as a noun, with the first cited use of the noun *design* occurring in 1588. *Merriam-Webster's* defines the noun as "a particular purpose held in view by an individual or group; deliberate, purposive planning; a mental project or scheme in which means to an end are laid down." Here, too, purpose and planning toward desired outcomes are central. Among these are "a preliminary sketch or outline showing the main features of something to be executed; an underlying scheme that governs functioning, developing or unfolding; a plan or protocol for carrying out or accomplishing something; the arrangement of elements or details in a product or work of art." Today, we design large, complex processes, systems, and services, and we design organizations and structures to produce them. Design has changed considerably since our remote ancestors made the first stone tools.

At a highly abstract level, Herbert Simon's definition covers nearly all imaginable instances of design. To design, Simon writes, is to "[devise] courses of action aimed at changing existing situations into preferred ones" (Simon, *The Sciences of the Artificial*, 2nd ed. [MIT Press, 1982], 129). Design, properly defined, is the entire process across the full range of domains required for any given outcome.

But the design process is always more than a general, abstract way of working. Design takes concrete form in the work of the service professions that meet human needs, a broad range of making and planning disciplines. These include industrial design, graphic design, textile design, furniture design, information design, process design, product design, interaction design, transportation design, educational design, systems design, urban design, design leadership, and design management, as well as architecture, engineering, information technology, and computer science.

These fields focus on different subjects and objects. They have distinct traditions, methods, and vocabularies, used and put into practice by distinct and often dissimilar professional groups. Although the traditions dividing these groups are distinct, common boundaries sometimes form a border. Where this happens, they serve as meeting points where common concerns build bridges. Today, ten challenges uniting the design professions form such a set of common concerns.

Three performance challenges, four substantive challenges, and three contextual challenges bind the design disciplines and professions together as a common field. The performance challenges arise because all design professions

1. act on the physical world,
2. address human needs, and
3. generate the built environment.

In the past, these common attributes were not sufficient to transcend the boundaries of tradition. Today, objective changes in the larger world give rise to four substantive challenges that are driving convergence in design practice and research. These substantive challenges are

1. increasingly ambiguous boundaries between artifacts, structure, and process;
2. increasingly large-scale social, economic, and industrial frames;
3. an increasingly complex environment of needs, requirements, and constraints; and
4. information content that often exceeds the value of physical substance.

These challenges require new frameworks of theory and research to address contemporary problem areas while solving specific cases and problems. In professional design practice, we often find that solving design problems requires interdisciplinary teams with a transdisciplinary focus. Fifty years ago, a sole practitioner and an assistant or two might have solved most design problems. Today, we need groups of people with skills across several disciplines and the additional skills that enable professionals to work with, listen to, and learn from each other as they solve problems.

Three contextual challenges define the nature of many design problems today. While many design problems function at a simpler level, these issues affect many of the major design problems that challenge us, and these challenges also affect simple design problems linked to complex social, mechanical, or technical systems. These issues are

1. a complex environment in which many projects or products cross the boundaries of several organizations, stakeholders, producers, and user groups;
2. projects or products that must meet the expectations of many organizations, stakeholders, producers, and users; and

3. demands at every level of production, distribution, reception, and control.

These ten challenges require a qualitatively different approach to professional design practice than was the case in earlier times. Past environments were simpler. They made simpler demands. Individual experience and personal development were sufficient for depth and substance in professional practice. While experience and development are still necessary, they are no longer sufficient. Most of today's design challenges require analytic and synthetic planning skills that cannot be developed through practice alone.

Professional design practice today involves advanced knowledge. This knowledge is not solely a higher level of professional practice. It is also a qualitatively different form of professional practice that emerges in response to the demands of the information society and the knowledge economy to which it gives rise.

In his essay "Why Design Education Must Change" (*Core77*, November 26, 2010), Donald Norman challenges the premises and practices of the design profession. In the past, designers operated on the belief that talent and a willingness to jump into problems with both feet give them an edge in solving problems. Norman writes:

> In the early days of industrial design, the work was primarily focused upon physical products. Today, however, designers work on organizational structure and social problems, on interaction, service, and experience design. Many problems involve complex social and political issues. As a result, designers have become applied behavioral scientists, but they are woefully undereducated for the task. Designers often fail to understand the complexity of the issues and the depth of knowledge already known. They claim that fresh eyes can produce novel solutions, but then they wonder why these solutions are seldom implemented, or if implemented, why they fail. Fresh eyes can indeed produce insightful results, but the eyes must also be educated and knowledgeable. Designers often lack the requisite understanding. Design schools do not train students about these complex issues, about the interlocking complexities of human and social behavior, about the behavioral sciences, technology, and business. There is little or no training in science, the scientific method, and experimental design.

This is not industrial design in the sense of designing products, but industry-related design, design as thought and action for solving problems and imagining new futures. This MIT Press series of books emphasizes strategic design to create value through innovative products and services, and it emphasizes design as service through rigorous creativity, critical inquiry, and an ethics of respectful design. This rests on a sense of understanding, empathy, and appreciation for people, for nature, and for the world we shape through design. Our goal as editors is to develop a series of vital conversations that help designers and researchers to serve business, industry, and the public sector for positive social and economic outcomes.

We will present books that bring a new sense of inquiry to the design, helping to shape a more reflective and stable design discipline able to support a stronger profession grounded in empirical research, generative concepts, and the solid theory that gives rise to what W. Edwards Deming described as profound knowledge (Deming, *The New Economics for Industry, Government, Education* [MIT Center for Advanced Engineering Study, 1993]). For Deming, a physicist, engineer, and designer, profound knowledge comprised systems thinking and the understanding of processes embedded in systems, an understanding of variation and the tools we need to understand variation, a theory of knowledge, and a foundation in human psychology. This is the beginning of "deep design"—the union of deep practice with robust intellectual inquiry.

A series on design thinking and theory faces the same challenges that we face as a profession. On one level, design is a general human process that we use to understand and to shape our world. Nevertheless, we cannot address this process or the world in its general, abstract form. Rather, we meet the challenges of design in specific challenges, addressing problems or ideas in a situated context. The challenges we face as designers today are as diverse as the problems clients bring us. We are involved in design for economic anchors, economic continuity, and economic growth. We design for urban needs and rural needs, for social development and creative communities. We are involved with environmental sustainability and economic policy, agriculture competitive crafts for export, competitive products and brands for micro-enterprises, developing new products for bottom-of-pyramid markets and redeveloping old products for mature or wealthy markets. Within the framework of design, we are also challenged to design for extreme situations; for biotech, nanotech, and new materials; for social business; as well as for conceptual challenges for worlds that do not yet exist (such as the world beyond the Kurzweil singularity) and for new visions of the world that does exist.

The Design Thinking, Design Theory series from the MIT Press will explore these issues and more—meeting them, examining them, and helping designers to address them.

Join us in this journey.

Ken Friedman   Erik Stolterman

Editors, Design Thinking, Design Theory Series

1 Making and Breaking the World

Enough Is Never Enough

Never have capitalist societies owned so much, wanted so much, and wasted so much. In a world becoming increasingly suffocated by people and possessions, we must ask ourselves what—beyond a conventional understanding of utility—is all this meaningful stuff really for, and why does it transform into meaningless rubbish so quickly? Slews of fully functioning products are committed to landfill each day. These landfill sites are not graveyards for dead products, as is often claimed; they are orphanages. Most products that end up there still function perfectly well but were simply discarded before their time.

In *Design for the Real World* (1971), the designer and educator Victor Papanek forcefully argued for a deeper ethics in design practice. At a time when rates of consumption and requisite environmental decline were skyrocketing, Papanek's writings necessarily drew our attention to the maladapted motivations of designers in the late 1970s. As an advocate of socially and ecologically responsible design, he saw the power of design in shaping positive change in the world and passionately advocated for it. Throughout his life and career, Papanek grew increasingly concerned about design's growing preoccupation with commercial success, often at the cost of environmental and social balance. He describes how for the first time in history, people have "sat down and seriously designed electric hairbrushes, rhinestone-covered file boxes, and mink carpeting for bathrooms, and then drawn up elaborate plans to make and sell these gadgets to millions of people."[1] Clearly, Papanek was ahead of his time—so much so that his words still ring true almost fifty year later. He lamented design's tendency toward morally and ethically bankrupt practices that make economic gain the key measure of creative success: if it sells well, it is successful; if it does not, it is not and so on. To Papanek, design was becoming nothing more than a slave to industry, instead of a powerful humanitarian tool with which to leverage positive social and environmental transformations. Of course, design's awakening social and environmental consciousness can also be traced back to Papanek's early writings and teachings. Although he cannot be held solely responsible, it was his voice that so many of us heard; it was his voice that sounded the call to action that so many of us responded to; and it was his voice that motivated generations of designers to dedicate their lives and careers to enabling design-led systems-level change.

Most resources taken out of the ground today become waste within only three months: waste comprising complex cocktails of plastics, metals, and other synthetic compounds no longer recognizable to Earth's microbial decomposers that degrade substances back to their basic nutritional building blocks. In an article for *Vice* titled "Future Relics," Caroline Haskins slays Apple's AirPods for their short-term, ecologically destructive character. Haskins, a senior technology reporter at BuzzFeed News, laments AirPods' eighteen-month life span, the way the lithium-ion batteries quickly stop holding charge and become unusable, and their inability to be repaired or recycled because they are glued together.[2] Products such as these are not unique; think single-use plastic water bottles or nonrechargeable batteries, for example. Products like these are symptomatic of a design mind-set that fails to grasp the systemic complexity of earth's dynamic, living systems. We know this because dropping products such as these into the world is like dropping a spoonful of arsenic into a koi pond: you just would not do it if you understood and cared about its ecological implications. As Haskins concludes, on a global scale, our capitalist system is predicated on a disregard for longevity, because companies consider it more profitable to make products that die than to make products that last.[3]

Socially approved patterns of behavior shape the way we interact with the material world. A surprising number of such norms govern the way we value and use goods. Many norms are so inherent that we are unaware of their existence. Norms are cultural products, including values, customs, and traditions,[4] which represent individuals' basic knowledge of what others do and think that they should do.[5] Sociologists describe norms as informal understandings that govern individuals' behavior in society. It is very normal, for example, that this book is written in English, though this might not have occurred to you until just now, when I mentioned it. Norms work in this way. They go unseen until someone draws attention to them, making them noticeable. Electricity works like this, too. Those of us in this world lucky enough to have electricity now find it so normal that we only really notice it when it fails, flickers, or cuts out entirely. The information science professor Susan Leigh Star articulates this point beautifully when she describes the way in which system failure bring the system itself to our attention, how failure reveals the previously concealed norm. She describes how "the normally invisible quality of working infrastructure becomes visible when it fails: the server is down, the bridge washes out, there is a power blackout. Even when there are backup mechanisms or procedures, their existence further highlights the now visible infrastructure."[6]

Norms also influence our perception of product life spans and how long our possessions ought to last; a toaster should last four years, a kettle five, and so on. None of us ever consciously decided this, but consumers were socially conditioned into these norms through the established behavioral patterns that preceded. They are massively influential in our decision-making processes relating to when a product should be replaced, or if it breaks, whether we fix it or dump it. Yet, despite their prominence, norms can often be quite ridiculous. For example, replacing a smartphone because of a cracked screen makes no sense, yet many modern consumers do this; their privilege blinds them to the reckless, idiotic nature of the act. From a materials efficiency perspective, doing this is about as ridiculous as replacing an entire car because of a flat tire. Both are equally repairable, yet one a quite common practice, the other unthinkable.

Modern consumers occupy what Walter Stahel refers to as a "production-oriented, fast-replacement system."[7] In 1982, Stahel founded the Product-Life Institute in Geneva to explore more resource-efficient strategies for economic growth. Since the mid-1970s, he has forcefully advocated for extending the service life of goods through reuse, repair, remanufacture, and upgrade. The frenetic state of the production-oriented, fast-replacement system that Stahel describes has "innovative products" romp into our lives and offer only the thinnest sliver of meaning before being thoughtlessly jettisoned as waste. We see this push toward innovation for innovation's sake most glaringly in the internet of things (IoT), in which nauseatingly shallow treatments are commonly assigned to technological affordances, undergirded by the misguided assumption that all people want is an easier life. This is simply wrong. Modern consumers do not just want easier lives; they want better lives, more fulfilling lives, and more meaningful lives. Indeed, the breathless pursuit of speed and convenience is something of a dead end. At this point, one might reflect on inconvenient, time-consuming things such as pets, home cooking, houseplants, friendships, parties, and travel. These things are certainly not "easy" or "convenient," yet they clearly have meaning and value to us. In "Spimes Not Things," Michael Stead, Paul Coulton, and Joseph Lindley lament design's current distraction with IoT, stating that "under a façade of innovation, the IoT is a breeding ground for superfluous, novelty gizmo products whose design incorporates environmentally damaging modes of manufacture, consumption and disposal."[8] Through their design manifesto for a sustainable IoT, these three design academics from Lancaster University lobby for a more thoughtful, less destructive mode of connected technologies—an IoT that nourishes and enables possible social and ecological futures, not one that destroys and shuts them down. Indeed, the giddy race toward a seamless world of integrated smart technology could actually turn out to be a race toward our species' demise. Yet connected technologies seem as though they are here to stay. In 2015 the number of "things" connected to the internet surpassed the number of people alive in the world. In 2018 there were an estimated twenty-five billion connected devices, and by 2020 a projected fifty billion.[9] How much longer can the exponential growth of connected "stuff" continue? Clearly we need a more sustainable IoT paradigm in place of the culturally impoverished, technocentric model that prevails.

In *Designs for the Pluriverse*, Arturo Escobar, a prominent Colombian American anthropologist who explores design-led visions for postdevelopment futures, exposes a paradoxical crisis underlying our feverish consumption of electronic products. He states: "So you are holding a digital device in your hand, maybe even while you read these pages. Do you know what it is? How it un/does you in particular ways? How it un/does the world? A narrowing in the diversity of existential options and choices open to us humans, the multiple ways of being in spaces and places in time, and of what technologies do to the earth and to our communities."[10] According to Escobar, this design-led narrowing of options is paradoxical, in that it is essentially propagated by digital artifacts sold to us via the promise of freedom. This, in many ways, builds on Tony Fry's premise of "defuturing."[11] Fry, a design theorist and philosopher exploring the relationship between design, unsustainability, and politics, denotes a closing down of future possibilities (defuturing), caused by design choices. Design, in this way, simultaneously opens up and closes down possible futures as it progresses. In creating a particular future, we inadvertently close others down. The sustainability crisis, therefore, is the aggregation of our species' myriad defuturing choices, when experienced in their inseparable, overwhelming totality.

Material consumption is out of control. Modern society witnesses an unprecedented acceleration of social life, characterized by increasing product replacement rates and short product life spans. In "Prometheus of the Everyday," the prominent Italian design professor Ezio Manzini forcefully argues that "the lack of a design culture capable of confronting these new technological possibilities has resulted in the dissemination of worthless products. So, the potential of the old technology is distributed in the banal forms of gadgets, disposable products, and ephemeral objects lacking any cultural significance."[12] He goes on to tell us how "a feeling of superficiality and the loss of relations with objects derives from this; we tend to perceive a disposable world: a world of objects without depth that leaves no trace in our memories, but does leave a growing mountain of refuse."[13] Manzini's work on design for sustainability has shaped much of the discourse in this space, providing solid theoretical foundations on which future generations of design theorists may continue to build. His work, in this way, is an accelerant of our design-led transition to more sustainable processes and practices.

Rapidly increasing replacement rates create social pressure to stay up to date and keep pace with technological development, resulting in harried and exhausted consumers.[14] In today's hyperconnected world, social media have fabricated a reality where everyone else seems fulfilled, ecstatic, and successful; everyone, that is, other than you. In the global North, we live in the age of envy: "Career envy, kitchen envy, children envy, food envy, upper arm envy, holiday envy. You name it, there's an envy for it."[15] As social animals, we humans have an innate tendency to compare, to evaluate our own success and failure in relation to those around us. Because we are a communal species, this is a natural thing to do, and the notion of how well you are doing only seems to make sense when set against the accomplishments of others. The clinical psychologist Rachel Andrew says that our use of platforms such as Facebook, Twitter, and Instagram elevates this deeply problematic psychological discord. Describing the toxic effect of social media on child and adolescent mental health, she likens it to "taking regular human comparing behaviours and putting them on steroids."[16] In this antagonistic culture of comparing, the feeling of "not having" is a common antecedent to consumptive behavior. The British sociologist Robert Bocock writes about the social psychology of material consumption. He describes how it is not that consumers always want what they buy; they just want to be liberated from the ache of not having it when everyone else around them does, a feeling he describes as a sense of "lack."[17] In this way, consumption is a competitive, status-seeking social practice. People buy things because of what they can do with them, what they can tell others about them, and what having them says about themselves. A product or service, therefore, is powerful because of how it connects people to something—or someone—else.[18] This principle holds true whether the product is physical, digital, or a hybrid of the two.

Regardless of an object's material or immaterial "form," every object originates from forests, soil, rocks, gas, and creatures; everything comes from the biosphere.[19] Through centuries of unchecked economic production and growth, capitalist societies have converted the organic, living systems we evolved within over eons into a pile of processed material artifacts, the majority of which now sit unused in attics, basements, or landfill. On average, each household in the United Kingdom has two old phones stored away somewhere—in other words, forty-nine million handsets simply sitting in our homes.[20] Fueled by an unquenchable thirst for innovation, consumers stride forth in pursuit of faster, lighter, stronger futures. In doing so, we leave a trail of discarded products behind us; toasters, flat-screen TVs, potato peelers, tennis shoes, wardrobes, and a host of other newly orphaned products fill the rear view. This wasteful process is exacerbated by our deployment of material goods to mediate our constantly changing identities. Once loved products fall out of favor all too quickly as they soon fail to keep up with the ever-evolving self.

Hedonic Adaptation

Why don't things last? Why is it that we acclimatize to things so quickly, and the new does not stay new for long? It often seems as though our desire for novelty, change, and difference continually grows, and the satiation of desire is only ever short-lived. In *The Overspent American*, the economist Juliet Schor describes the "upward creep of desire,"[21] the slow but sure ascendancy of our want for new experiences, fueled by our innate ability to outgrow and overcome. Much of Schor's work traces clear pathways that connect social anxiety with material consumption, and material consumption with climate change. Her important work thus lobbies for a deeper questioning of our everyday social practices, if we are to tackle larger-scale problems like the collapse of the biosphere. Of course, identity seeking can play a positive role as we explore ourselves through the process of material consumption. It is through this process that material things help us make, and remake, ourselves. We write our own story with material things, to a greater extent than is often assumed. This story is continually unfolding, being updated, and retold. As our circumstances change, so too does our story. Much of Schor's work points to status-seeking behaviors and our desire to feel affiliated with communities, ideologies, and lifestyles. She balks at how some people "have so little sense of themselves that they buy a shirt emblazoned with Tommy Hilfiger's phony coat of arms or buy desert-ready Land Rovers to commute over asphalt."[22] This display mode of conspicuous consumption deploys material goods as signifiers of status, albeit weak signifiers. It has us assembling a pageant of material things and has us chasing cars, jackets, houses, and shoes to provide props for this performance. The "upward creep" Schor refers to is very real, at times to the extent that it feels as though it were in our bones. Everyday life in affluent societies is characterized by a kind of hedonism that fuels the continual pursuit of more. But the relevant move here is to ask *why* we are like that, rather than to state *that* we are like that.

"Hedonic adaptation" is the observed tendency of humans to quickly return to a relatively stable level of happiness despite major positive or negative events or life changes.[23] Sometimes referred to as the "hedonic treadmill," this idea is described by many positive psychologists as a process that reduces the affective impact of emotional events. According to Sonja Lyubomirsky, a professor of psychology at the University of California, Riverside, who specializes in personality psychology and the pursuit of happiness, empirical and anecdotal evidence for hedonic adaptation suggests that both positive and negative experiences fade with time, returning eventually to a kind of experiential baseline. She states that "the thrill of victory and the agony of defeat abate with time. So do the pleasure of a new sports car, the despondency after a failed romance, the delight over a job offer, and the distress of a painful diagnosis."[24] Like a bungee cord pulling us back to a point of origin, hedonic adaptation is the tension in that cord, the restorative force drawing us reliably back to our default state. This is not to say that people are incapable of change or that this emotional baseline is indelibly fixed or immovable. Rather, our emotional states are not as subject to long-term change as we have been led to believe. Our pursuit of this mythology, through the feverish consumption of material things, has now taken us to the brink of ecological collapse. Fyodor Dostoyevsky proposed that the human "is a pliant animal; a being who gets accustomed to anything."[25] According to Lyubomirsky, people adapt more quickly to positive experiences. Evidence has shown that the impact of everyday negative events is more powerful and longer lasting than that of positive events.[26] We reflect less on positive events, and so they are less deeply embedded. Ordinarily, we do not relive positive events in our mind to the same extent as negative events. Positive events are thus less deeply embedded and as such are easier to overcome. This embedding is like the way we might revise for an exam. We "relive" the information repeatedly to more firmly embed it within our long-term memory system. The Canadian psychology professor Gerald Cupchik describes this predictable fading of emotional affect as "the half-life of an emotion."[27] Cupchik posits that emotional reactions emerge in what he calls a "perfect storm," whereby meaningful situations evoke bodily memories that unconsciously shape and unify the experience.[28] In giving emotional affect a kind of half-life, the process of hedonic adaptation limits the severity and duration of the blows and traumas people receive in life.[29] The social psychologists Timothy Wilson and Daniel Gilbert aptly nickname this process a "psychological immune system" that helps us deal with life's unpredictable ups and downs.[30] They build on this idea through what they call "affective forecasting," which is the process of predicting how you will feel in the future, and the subsequent actions you take to try to shape that future state.

From an animalistic perspective, hedonic adaptation has an important survival component. For example, it helps explain why people tend to recover from both positive and negative life events.[31] For our psychological well-being, this is probably just as well. Imagine if you experienced rage over a parking ticket, and this rage never faded. Or if you experienced sexual arousal, and this heightened state became your new experiential reality. We would not be able to function if this were the case. If emotional affect held fast and never faded, we would quickly become psychologically overwhelmed. And so emotional experiences erode over time as we adapt to them. This phenomenon is frequently observed in lottery winners, who report years later that they feel no happier than they were before the influx of wealth. Many even speak of a reduced state of well-being, once the initial novelty of being wealthy had worn off. The increase in wealth led to no permanent increase in happiness, at least not to the level one might expect.

The power of personal possessions as designators of our being cannot be overstated. Much of the discourse emerging from the field of material culture studies knows this, having taken this position for well over a century. Material culture is the dimension of social reality that is grounded in the materials, objects, and spaces that collectively form the designed world. These assertions are often founded on an "additive story" that argues how acquiring things takes you from here to there—from a current state to a preferred one. Much of this work emerges from the consumer psychology perspective and seeks to uncover "why we buy this and not that" or "why we see meaning in these things and not those things." Of course, the capitalist context from which this work emerges is clear and focuses primarily on cognitive processes of "wanting and getting," rather than "having and keeping." The simplistic ideology of much material consumption is "if I buy *this*, then I will become more like *that*." In this way, we consume our way toward a preferred state while simultaneously consuming our way away from an unwanted state. We consume not only "things" but also the promises they mediate: how they will lift our mood, reduce our sense of inadequacy, makes us feel more up to date, and so on. Though these promises are always made, and sometimes even delivered, they are seldom kept. We outgrow their novelty quickly and soon become distracted by newer models, making newer and bigger promises. In this way, hedonic consumption is deeply attractive because of the immediate sensory gratification it offers.[32] After one interacts with such a product, if it no longer captures one's attention or elicits emotional affect, one can be said to have adapted to it.[33] With this in mind, we should not be surprised that a product that promises excitement from the outset will soon fail to deliver excitement. Unless products themselves can adapt and change with us, we will outgrow them quickly as they fail to keep up. Their predictability, like rot, inevitably sets in.

Valuable research on "stripping" adds further support to the assertion that our possessions—be they material, digital, or otherwise—powerfully anchor us in our own constructed realities. Stripping, in this sense, refers to the involuntary removal of an individual's personal effects as a form of imposition, punishment, and control designed to disempower incarcerated people by stripping them of their identities. This stripping process is commonly seen when individuals are incarcerated or institutionalized in places like prisons or mental hospitals. Melanie Wallendorf and Eric Arnould, professors of marketing and consumer culture, argue vehemently that objects carry a self-concept-based meaning and are far more experientially complex assemblages than is conventionally assumed. Losing or severing our connection to material possessions can involuntarily change the very meaning of life for individuals. Wallendorf and Arnould describe the point of incarceration, in which "an individual's clothing and personal possessions are taken away. Institutional clothing and objects are issued for the person's use but are not under his or her full control. Thus, ownership of objects disappears as the institution takes on the role of providing objects for one's use. Connections to 'normal' life on the 'outside' are severed, and individuals gradually assume the dependent role of patient or prisoner."[34] This phenomenon plays out when elderly people move into a nursing home: they experience a loss of status and quickly feel unmoored and lost. Cherished objects like photographs of loved ones or ornaments that elicit meaningful associations are brought along to help fill the void and reestablish that lost sense of identity and belonging. We also see this in cases of dementia, where patients suffering from severe memory loss also experience a requisite loss of their identities. In this way, we might consider a person's memories as a collection of possessions, gathered over time, which collectively establish a sense of identity and place.

Unlike relationships with people, product relations are relatively limited in terms of their adaptive, evolutionary capabilities. Products, in this way, are relatively static and fixed. They are easily outgrown, and as such, their novelty tends to fade quickly. In contrast to the fleeting nature of emotional affect, the "meaningful associations" we form with designed objects can be far more long-lasting, even permanent. And so when we speak of wastefulness, and of products that do not last, it is important to remain at least partially aware of our evolutionary and animalistic predisposition, of our innate ability to overcome, to outgrow and to adapt to novelty. For design, it is essential to recognize this and to accept that it happens because of our nature. Indeed, designing a rich "emotional experience" is one thing, but emotions do not last and were never meant to. The things that last are the meaningful associations we form with objects over time. These deeper experiences do have the potential to endure, and to take root in the hearts and minds of users. And so, if we are to have a meaningful discussion about product experience, we must do so longitudinally. Product experience happens throughout a product's life, often spanning years, even decades. Emotional affect plays a part in the curation of rich product experiences, but only a part. Emotions fade; they simply do not last. Thus, for user experience designers to hinge a long-lasting product relationship on emotion alone is pure folly; it is really all about meaning.

A perpetual state of dissatisfaction characterizes the human condition. To a degree, this is in our biological makeup. In today's world of hyperconsumption, enough is never enough. The much-sought-after experience of being up to date is fleeting. Our choice to engage with designed objects is motivated by many parallel motives. Through encountering objects, this complex and dynamic assemblage of motives coalesces to shape the nature of our interaction. And so, to say that the consumption of material things is motivated by either "this" or "that" is reductive to the point of distortion. Yes, it is possible to singularly see modern consumers as nothing more than docile sheep, marching to the steady beat of a marketer's drum, or puppets with strings tethered to the hands of a capitalist controller. Yet, despite the prominence of these forms of rhetoric, they offer a lazy critique that fails to capture the multifaceted character of consumption as a dynamic social process. Such a perspective also deeply underestimates the complexity of our multilevel and highly nuanced interactions with the designed world. Indeed, as a designer, you subscribe to such overly simplistic and pejorative explanations at your own peril. Much academic writing is devoted to this kind of "mudslinging," which dismantles the human into a collection of superficial and self-serving parts. This is both wrong and unhelpful, just as we are not all ungrateful sinners, as some books would have us believe. We all have the capacity to do wrong, sure, but this capacity lives side by side with many other capacities, such as our capacity to care for others, for one. Such defeatist narratives paint the bleakest of pictures and show us at our very worst. It is true that consumption can be a process through which people curate stories about themselves, to themselves and others. But this is not necessarily a bad thing, and it is certainly a skin-deep analysis that claims as such. Who is to say what is superficial and what is not, what is necessary and what is not, and what matters and what does not? It is not helpful to do so. What is more helpful is to notice that people commonly do these things, and to work harder and more thoughtfully in working to understand why. After all, the underpinning motives and drivers shaping these behaviors are of far greater value to design than the surface-level behaviors they give rise to.

Designed to Fail

The deliberate curtailment of product life spans to sell more stuff is surprisingly common practice in all manufacturing industries. For some producers, their business model was designed around this policy, and so to change feels almost unthinkable, as it has come to characterize the very basis of their survival. A number of these companies are harming themselves through this practice. Customers are not getting products that can perform adequately and safely for a reasonable amount of time, and the result is proving to be more detrimental than beneficial.[35] Clearly, short-life products are a powerful accelerant of short-term consumption and waste, but an equally powerful corruptor of the long-term credibility and value of the businesses that make such products.

The apparent decline in certain product lifetimes has been attributed to pressure on manufacturers to maintain sales, leading them to introduce stylistic changes to make older products appear "out of date" and reduce product prices by making minor, incremental reductions in quality.[36] Take the streamlining craze in the 1940s, for example. This "stimulated a wealth of new products for which the style was functionally unnecessary or even wildly inappropriate: radios, electric heaters, vacuum cleaners, irons, toasters, jugs, pans, light fittings, cash registers, even staple guns and a pencil sharpener."[37] Stylistic design movements such as streamlining were deployed as accelerants to consumption—ways to get people spending. The design historian Alice Twemlow laments the increasingly commercial orientation of design. She describes how "in the US in the 1950s the discussion of Design was linked to efforts to boost the sales of American industrial design in the market place. Such discussion was led, for the most part, not by government bodies, however, but by an alliance of cultural institutions, publishing, and retail businesses."[38] New generations of designers dived headlong into this new, aggressively commercial space, as did the consumers who flocked around the glossy, aerodynamic forms those designers created. In the mid-1940s, when asked for his thoughts about aesthetics in design, the French-born American industrial designer Raymond Loewy described how it "consists of a beautiful sales curve, shooting upward."[39] Clearly, this witty remark was made at a time when the connections between consumption and environmental decline were relatively underexplored. Known as the father of streamlining, Loewy with this remark brazenly attests to the long-standing relationship between tweaks in style and the artificial stimulation of consumer desire—a wasteful and destructive coupling that we remain saddled with to this day.

Planned obsolescence involves a design plan that is intended to hasten existing products to become undesirable either functionally or psychologically, and consequently to be replaced by newer products.[40] In the case of "functional obsolescence," we find consumer products no longer able to perform the task for which they were originally created. This departure from functionality is most commonly caused by objective factors like acceleration (e.g., product is slow relative to newer models), compatibility (e.g., format not supported by newer system), or infrastructure (e.g., spare parts no longer available). With "psychological obsolescence," on the other hand, product value is vulnerable to several subjective factors. These most commonly include trend (e.g., tastes shift, rendering the product stylistically out of date), identity (e.g., products no longer reflect your desired identity or represent you in a favorable way), and meaning (e.g., we form unwanted meaningful associations with products, leading to their eviction from our lives).

Indeed, things do not last as long as they used to, but the notion of a throwaway society is nothing new. Bernard London, an American real estate broker, first introduced the term "planned obsolescence" in 1932—popularized by the acclaimed American journalist and critic of consumerism Vance Packard in his book *The Waste Makers* (1964)—to stimulate spending among the few consumers who had disposable income during the Great Depression (1929-1938). In fact, London wrote several essays around that time (1930-1935) contending that increased levels of material consumption, stimulated by making people feel dissatisfied with the things they already own, offered the most direct path to economic growth.

In the United States, "The Great Depression ended around 1939, but London's idea, it seems, lives on. Today, mass consumption and shopping have become a way of life; consumers sometimes do not even wait for something to break or wear down before replacing them."[41] By the late 1950s, planned obsolescence established itself as the dominant economic paradigm, stimulating swaths of manufacturers to design short-life goods no longer built to last. In the United States of the early 1900s, when people referred to something as being "disposable," they would most likely be discussing cheap, materially lean objects like diapers, newspapers, or tin cans. Today it appears that everything is disposable—from a barely used hair dryer, raincoat, or vacuum cleaner to an entire fitted bathroom or kitchen. One might even wonder whether this materially oriented short-termism holds some influence over our interpersonal relations too, as many of us become increasingly promiscuous and ephemeral in our social relations. Each of us claims to have a greater number of "friends" than ever before, but seldom are these frequent, abundant friendships built to last.

And so planned obsolesce is not a new phenomenon by any means. "The idea that one disposes of artefacts or products before one actually needs to in order to buy a more up-to-date or desirable version is at least as old as consumerism, and capitalist society itself."[42] But to knowingly design a product so that it quickly fails—either on functional or psychological grounds—is a more recent development. Even in the packaging of consumables, it is well known that brands create containers that promote excess use and the discarding of unused product. Think of the "empty" ketchup bottle with the two tablespoons worth of sauce plastered to its plastic interior walls, or the scrunched-up discarded toothpaste tube with a week's supply of paste trapped within its crumpled polymer shell. Industry designs products to artificially accelerate the rate of consumption of olive oil, perfumes, sunscreens, moisturizing creams, shampoos, and other related products. In the case of the toothpaste tube, for example, increasing the internal diameter of the opening from 5 mm to 8 mm boosts rate of consumption by about 60 percent.[43] In the case of software, we see "upward compatibility only" as a strategy to accelerate technical obsolescence: bringing forward to the upgrade point by enabling the new version of a program to read files generated by older versions, but not the other way around. Even disposable electronic products are commonplace today: abhorrent gadgets like digital wine thermometers with batteries that cannot be replaced once they have run down, or key chain flashlights with the same idiotic disposability. In the summer of 2019, it was estimated that Britons bought fifty million throwaway outfits.[44] These cheap, single-use garments are worn once or twice on holiday and then dumped to avoid carrying them home and washing them. I once saw a single-use phone-charging battery for people on the move. The advertisement for these "emergency chargers" celebrates their ease of use, describing how you simply tear off the plastic packaging, plug the 1300 mAh battery into your smartphone, and, once it transfers all its charge, simply throw it away—voilà! This kind of morally bankrupt industrial design is alarmingly common and shows how the true depth of our stupidity frequently comes clothed in the sleek disguise of "innovation."

Of course, products frequently fail of their own accord and probably always will. For example, "Latches on pencil cases aren't meant to fail after some number of years or some thousands of operational cycles; they just fail. It is the same with laptop keys or paper clips or duct tape."[45] In these cases, the failure was not planned but an unintended side effect of designing a product within certain constraints (e.g., materials availability, price point, market demand, client-side requirements). Not all obsolescence is planned; think hard copy encyclopedias or floppy disks, for example. It was never the "design plan" for these typologies of object to befall disuse in the way they have. Rather, they testify to a form of unplanned obsolescence that emerges in response to the continual unfolding of technology.

Gaia Vince, an award-winning science journalist, describes her frustration with how consumer products are not designed to last. She recounts the ordeal of her broken camera: "My camera shutter, battered by the dust and grime of travel, no longer works. I'm told I should throw away my camera, even though it works fine, apart from the shutter mechanism."[46] Vince became the first woman to win the Royal Society Winton Prize for Science Books with her monograph *Adventures in the Anthropocene* (2015). Eyewitness accounts of product failure—like Vince's account of her broken camera—are commonplace in today's world of mass-produced objects that are designed to be sold, briefly used, and replaced at the first sign of failure. A host of such tactics are commonly deployed with the intent of downstream failure. In the case of white goods like washing machines, it might be that a silicone seal around the hose connectors, or a metal bearing inside the electric motor, is lower grade and destined to fail after only a short period of light use. These small components are often more expensive to repair than the cost of replacing the entire machine. Similarly, "Specific lifespans are programmed by the manufacturers into chips in some equipment, so that printers will stop working after a pre-set number of pages, coffee makers will cease functioning after a pre-set quota of brews, and memory cards will stop uploading after a pre-set number of photo uploads."[47] Whether planned or not, the Apple Watch has obsolescence woven throughout. According to the tech repair and upgrade website iFixit, the watch's S1 SiP (internal system-in-package) is encased in resin and further held in place by a mess of glue and soldered ribbon connectors, making component replacement impossible. Though environmentally problematic, products like the Apple Watch are still important. For example, a friend of mine, Marilyn, is in her eighties and loves hers. It helps her feel safe and reachable and alerts close family if she were to have a fall—all without having a clumsy piece of bulky, beige medical equipment strapped to her arm, reminding her every minute of every day that she is infirm and potentially moments from her next calamity.

In a "smart" and increasingly connected world, products like this are commonplace. Yet, despite their brief use careers, they are built never to be dismantled. They combine plastics and metals in a form that can only follow a linear path from earth to earth—from mines and oil wells, through a brief visit to someone's home, back to a landfill site. Or perhaps the polymer case will degrade and find its way into groundwater flows, then be flushed out to sea—a biome that by 2050 will contain more plastic than fish. When products are designed for limited use and complexity of repair, consumers are forced into wasteful discarding and replacement behaviors by simple economics. This kind of obsolescence consists of a form of "design plan" that hastens the product in becoming dysfunctional or undesirable and, in so doing, makes it easier for the user to discard and replace it. In accelerating consumption and waste in this way, we also accelerate our destruction of the ecological systems that support life on our planet.

Matter, Flowing Through

In terms of design's responsibility, we have to "own" the things we design and make, while also "owning" the unforeseen social and environmental impacts of those things. That, essentially, is the kind of thinking that sustainable design calls for. Indeed, designing a product and not caring about the social or ecological impact of that product is just silly at this point. Despite being an incredibly dynamic and vibrant cultural phenomenon, design can also be an extremely wasteful and destructive one. This is largely due to its ephemeral nature, fueled by the ceaseless consumer hunt for change, novelty, and innovation. This problem calls for a new form of sustainable design thinking. By this, I am certainly not referring to the fast-tracked mode of "design thinking" that has you attend workshops and play designer for a day, where well-meaning gangs of adults giddily exfoliate several bricks of sticky notes in the name of innovation. Imagine if a bunch of designers wanted to play "pharmacist" for a day and spent a few hours dressed in lab coats, randomly mixing stuff together to see what happens. This is a clumsy caricature of design and nothing more. No, this is the opposite of what I mean by "a new form of sustainable design thinking." Design is an opportunist, adaptive process of continual development, innovation, and emergence. This persistent evolution responds to shifts in social, cultural, technological, and economic norms and trends and is unrelenting in its forward thrust. Despite this seemingly progressive character, design's recent transition from a "world-making" to a "world-breaking" enterprise has thrown it into flux; reexamination of the potential of designers as agents of positive change must continue to gather in intensity.

In *Creed or Chaos* (1947), Dorothy L. Sayers frames the crisis with language that, for me, conjures palpable sensations of disgust and shame. She writes how "a great curse of gluttony" has shaped our overproduction and overconsumption, a curse that inevitably concludes with our "destroying all sense of the precious, the unique, the irreplaceable." Sayers bemoans the profundity of waste overspilling our material worlds through pointless excess and an insatiable desire for more; she describes these unwanted material things as "the slop and swill that pour down the sewers over which the palace of gluttony is built."[48] If design is to meaningfully tackle this so-called consumer gluttony, we must do so across the full spectrum of material throughput: beginning, middle, and end. The endeavor, therefore, is to stem this flow, by reducing both consumption *and* waste, not one or the other. The discursive space connecting the two social practices provides the critical territory in which design, and this book, resides. The notion of "throughput" is ordinarily reserved for industrial processes: technical analyses of supply chain optimization or gross domestic product (GDP), for example. Aside from commercial processes, we can also consider the term useful. In the context of a person's material world, throughput describes the way material things "flow through" one's life. Throughout the book, I speak of "wastefulness" in several different ways. What each variant form shares, however, is a deep connection to the notion of throughput. In design, you cannot meaningfully talk about waste without simultaneously talking about consumption, and vice versa. Indeed, without such a holistic, system-level view, design thinking becomes naively reductive, unhelpfully fragmented, and grossly ineffective. We find ourselves looking at facets of the larger system and then wonder why our design interventions fail to enable the kinds of change we envision.

Though we in the global North produce more waste than ever before, it is important to note that "wastefulness," as a social practice, has not necessarily increased as much as one might assume. Martin O'Brien, reader in criminology at the University of Central Lancashire, writes eloquently about the sociology of waste. Through empirical observation of everyday consumption and waste practices, his research reveals previously concealed patterns and divisions across gender, class, race, and ethnicity. He cautions how "the temptation to view with alarm the apparent increase in the generation of waste is understandable but the alarm needs to be tempered with a realistic assessment of what lies behind it."[49] His point rails awkwardly against the populist left-wing rhetoric that lambastes modern consumers as ever more wasteful cretins, thoughtlessly wrecking the planet through conspicuous consumption. Frequently, this form of "negative critique is directed at post-war social development and implies that there is something peculiarly wasteful about contemporary society; that modern consumers are uniquely profligate, ignorant, and disdainful of their consumption behaviour compared to their parents and grandparents."[50] Of course, awkward truths lurk beneath this statement, and modern consumers do produce far greater quantities of waste than ever before. Moreover, the toxicity of today's waste is significantly greater than it ever has been. Yet we must also recognize that we also consume far more than ever before, so proportionally we might well be just as wasteful as we have always been. The ratio of material input (consumption) and output (waste) may not have changed as much as we might think. Think of filling a drinking glass from the kitchen tap: the faster the water from the tap flows, the faster the glass overflows. Slow down the tap, and the glass continues to overflow, but at a slower rate. The glass cannot be described as more or less wasteful than it was before; it is just responding in a consistent way to the rate of throughput, or to the speed at which material flows through it. Similarly, consumption and waste are corelated. There is more waste today, but there is also more consumption. Far from drawing a pedantic distinction, this reframes consumption and waste as "material throughput," flowing through our lives.

Meaning Is Emergent

In this book, I explore how meaning is proposed by designed things and then precariously held together through our ongoing interactions with them. This, therefore, is a book about interaction design; a book about the designed encounters through which people and things intertwine; and a book about how the longevity of our relationships with products is tied to the fate of our world. Indeed, keeping our "meaningful stuff" longer can make a significant difference in resource efficiency, more so than most conventionally approved sustainability practices. Indeed, "Whilst many people believe they are doing the right thing, and they most likely are, quite often we are having less impact than we think. For example, leaving chargers plugged into the power outlet is often held up as an example of a behavioral eco-crime. The truth is that the amount of energy saved by switching off a charger is the same as the energy used by driving an average car for one second."[51] Even the much-promoted practice of recycling is an ethical discarding practice. It is better than nothing, but in the end, it is still about discarding.

In terms of interaction design, "meaning" is the relationship that forms between the subject and the object. As an emergent property, meaning is the result of affective dialogue between people, phenomena, things, and one another, as encountered in the world. Meaning is fluid in this way and takes form as interactions between users and products unfold over time. In *Acts of Meaning*, the developmental psychologist Jerome Bruner defines meaning making as the construction of an individual's logic through interactions with people and things.[52] As with the meaning of words and phrases, so object meaning is highly unstable and continually adapting to its context. For example, an apple on a teacher's desk and an apple in a chef's pantry mean distinctly different things. Jean Aitchison, professor of language and communication at the University of Oxford, describes how "word meanings are like stretchy pullovers, whose outline contour is visible, but whose detailed shape varies with use."[53] Products, too, are like this. Their meaning is not static but dynamic and continually shifting in relation to the social and cultural environment.

While meaning is indeed a complex notion, there are essentially two types of meaning: "semantic" and "pragmatic."[54] Semantic meaning is concerned with the literal meaning of things, whereas pragmatic meaning focuses on the inferences and subtexts that context triggers in the mind of the perceiver. With semantic meaning, the meaning of the material, product, or service exists objectively without the context and may be defined as such. It is relatively stable and less vulnerable to individual factors such as one's age, taste regimes, and personal history. In contrast, pragmatic meaning considers how meaning changes in relation to its social and cultural context. It is highly unstable and wholly vulnerable to the idiosyncratic variances of each individual person. Both forms of meaning exist simultaneously, and it is the interplay between the semantic and the pragmatic that brings meaning to life. The study of psycholinguistics has well established that pragmatic meaning forms through the subjective inferring of nonliteral messages presented through speech.[55] Transferring this idea to the way users might infer meaning from their material things makes sense. At this point, we see that meaning is amorphous and plural. To this effect, Dmitry Leontiev differentiates between "the" meaning and "a" meaning.[56] This distinction helpfully detaches us from the idea that meaning exists as a singular truth that must simply be connected to, and replaces it with the more sophisticated idea that we are connecting to multiple meanings and truths, all at once. Meaning therefore exists within a dynamic ecosystem. It is adaptive, vibrant, and alive and describes the dynamic, living relationships between objects, phenomena, happenings, and actions that are part of one's life.[57]

The meaningful associations we form with the designed objects around us are vulnerable to a host of external influences (e.g., weather, time of day, personal preferences, what objects are placed next to). As we read a text or listen to the words of a friend, for example, we simultaneously coauthor a metanarrative in our minds, resituating the (semantic) meaning of our friend's words into our own context and, in so doing, inferring an additional layer of (pragmatic) meaning from them. The same process unfolds when we interact with products, and our experience of them is powerfully shaped by this coproduced flux of meaning, narrative, and metanarrative.

Though connected to the product, these metanarratives float somewhere above it, tethered by a thin thread of subjective interpretation, like a kite dancing in the wind. The meaningful associations formed by an encounter with an app, for example, tell us something about the app itself, but they also serve as residue of the cultural constructs that are appropriated to provide frameworks for thinking.[58] A person's cosmology, therefore, plays as much a part of the meaning-making process as the material, finish, or style of a given object. In this sense, each of us is a coproducer of meaning. Designed things simply provide the wireframe architecture on which we may drape layers of meaningful association over time.

2 Cultures of Keeping

From Product to Possession

We all own things that we are intimately connected to, things we treasure, things that hold tremendous personal meaning and significance beyond their utilitarian value. Often these possessions have little or no monetary value but are precious to us in other, more complex ways. Compared to the plethora of everyday objects that surround us, "cherished possessions" are rare, though each of us owns at least one or two. The stories behind them are rich and varied. Occasionally, these treasured artifacts have no utility value whatsoever but are kept for the meaningful associations they elicit—a watch given to you by a now deceased relative, a letter from an old friend, a splinter of doorframe from the house you grew up in. All these treasures connect us to times, places, and people of significance: these are the things that matter.

Meaningful possessions create constancy in a world of relentless change. The experiential gulf separating "anonymous products" from "cherished possessions" is vast. Within this nuanced difference, material things are unconsciously assigned value according to their use as performative tools (how well they work) and significance of symbolic value (their meaning to us). The formation of value is culturally influenced and always involves socially informed judgments of qualities such as goodness, worth, truth, justice, or beauty.[1] The German sociologist Max Weber defines two patterns of value creation: "instrumental" and "intrinsic." Weber was one of the founders of sociology as a field of inquiry, along with Émile Durkheim and Karl Marx. To Weber, "instrumental value" forms when something works well as a tool and performs the task it was designed to do (e.g., a disposable pen). In contrast, "intrinsic value" forms when something is good in and of itself, aside from its utilitarian purpose (e.g., a favorite pair of socks with holes worn through both heels).

Despite these two forms of value being distinct from each other, we frequently experience both forms simultaneously through a single object—just not always in equal measure. For example, an old washing machine might elicit large degrees of instrumental value (it washes the clothes well and does not take too long), but a small degree of intrinsic value (it is the first washing machine we owned, and I remember the day we bought it). According to Weber, instrumental value is objective. It is measured by how well a thing performs, and the extent to which it executes the task it was purchased for. This aligns with the Heideggerian notion of "equipmentality,"[2] whereby an object is a "something in-order-to." In this way, instrumental values are less emotive, as performativity can be impartially evaluated against universally held indicators. In contrast, intrinsic values are subjective. They provide an internal reference for what is good, beneficial, important, useful, beautiful, desirable, and constructive, for example. These subjective variables are strongly influenced by emotional factors and vary from person to person.[3] Of course, our subjectivity is shaped by the context we are in and have become accustomed to. As such, intrinsic values are socially influenced and powerfully shaped by the norms and practices of a given cultural context.

Amid the swarm of material things that encircle us every day, we share differing degrees of intimacy with each one. This so-called intimacy is not necessarily related to the frequency of use or how often we interact with a thing. For example, I use my mechanical pencil several times a day, but I do not feel any great personal connection to it. It is a nice pencil, but to me, it is just a product that works well. It has high instrumental value, but low intrinsic value. The meaning it has for me is probably like the meaning it would have for you. In fact, until I started describing it just now, I had never really thought about it. If it were to break, I would be a little irritated, but I would just buy another pencil. If I were unable to repair it myself, I would not bother getting it repaired professionally, because I do not "care" enough about it.

In contrast, there are other things I use only very occasionally, but which are incredibly important to me. Should one of those items get lost or break, I would invest significant amounts of my own time and money in either finding or repairing the item. I have a sashimi knife that I was given as a gift when I left my job as a chef several decades ago. The knife is beautiful, not just in the way it looks, but also in the way it works. A disproportionately long blade compared to the handle allows a sliver of fish to be sliced cleanly in one single stroke. It does need maintaining from time to time, and the ritual of honing the blade to restore its razor-sharp edge is a ritual I enjoy. I take my time over the sharpening process, which is usually accompanied by a strong cup of Italian coffee. The next time I use the knife, it feels even better than when it was new as it glides effortlessly through whatever it is cutting. I also feel a certain ownership of the experience, as it was I who invested time and energy to reinstate the blade's high performance. This small investment of care casts me in the role of cocreator, as though I played a small part in producing this extraordinary product. Sure, that the knife was a gift is central to its success; moreover, it is a gift that I like, and one that was thoughtfully chosen.

As the philosopher and sociologist Theodor Adorno describes, real gift giving has its joy in imagining the joy of the receiver. It means choosing, expending time, going out of one's way, thinking of the other as a subject.[4] Commonly recognized as one of Europe's most influential postwar philosophers, Adorno was central to establishing "critical social theory"—an approach commonly applied to the work of the Frankfurt school of which he was a part, aimed at questioning the oppressive political underpinnings of Marxist social criticism. Adorno laments the increasingly anonymous and thoughtless character of our gift giving. He refers to this as a violation of the exchange principle, in which gifts should demonstrate your understanding of the receiver, and what the receiver would like, rather than forcing one's preferences on another in material form. The gift of the sashimi knife also marked a key moment in my life. I was leaving a job, and a whole way of life, to go traveling in northern Japan. The journey changed the entire course of my life. And so the intrinsic value of the object was already high for me, regardless of its instrumental value and how well it performed as a cutting tool. Perhaps it is also because I do not use it too often that the knife remains special. My use of it is "rationed," and so I have not become overfamiliar with it.

For many years, I have been running "object handling sessions." In these sessions, people come together in a group to discuss their most cherished possessions. The process operates as an ongoing series of focus groups, with personal possessions providing the locus for individual testimony and group discussion. This form of "narrative inquiry" uses artifacts and life experience as the units of analysis to research and understand the way people create meaning in their lives.[5] These sessions are fueled by my own personal and professional curiosities about the reasons why we keep certain things and let go of others. They provide a personal and intimate means of having this kind of discussion, separate from oft-anemic theoretical debates and emotionally detached academic discourses. These conversations allow me to better understand the "lived experience" of relating to material things and have further expanded my understanding of the experiential phenomena that shape patterns of consumption and waste.

Through these sessions, I have found that each of us owns things we are either weakly or strongly attached to. Often we are unaware of the significance of these things until we really think about it. It is not until someone asks us why we cherish an object that we are able to explore and reflect on the reasons why it matters so much to us. We can sit and reflect within ourselves on these reasons, but it is through a conversation with an interested listener that the nuanced detail of these object stories rises to the surface. Like pulling on a loose thread, retelling these object stories unravels rich histories and intricate personal narratives. From the listeners' points of view, there is something powerful about hearing the stories of others. It helps us reflect on our own stories and retell them with more detail and granularity. I find these sessions fascinating, and I never fail to be surprised by the diversity of material things that successfully make the transition from product to possession.

Regardless of who participates in a session and how experienced they might be in reflecting on these kinds of matters, sessions always follow the same format. A typical object handling session lasts ninety minutes and involves ten to fifteen participants. One week before each session, participants are asked to choose their most cherished possession, bring it to the session, and be ready to talk about why it matters to them. Once at the event, I introduce the session so that people understand what is happening, but I also try not to say too much, as I do not want to influence the conversations that will emerge. Then all the objects are displayed in a neat group on a clean surface. We look at the objects together, in silence. Participants are always curious to see what others chose to bring, and cannot help but wonder why each object might hold such an important place in a person's life. I always bring a cherished possession of my own to talk about, which I place anonymously within the collection of other objects. I go first, by picking up my object and talking about what it is and why it matters. After telling my object story, I put it down and point to another object in the group. Whoever brought that object now takes his or her turn to introduce the object to the group and describe its relevance. Once finished, that person then chooses the next object, and so on, until all the objects have been discussed. The session concludes with a general discussion of any patterns, themes, or insights emerging from the discussions.

Over the years, I have run well over a hundred of these sessions all over the world. I have run sessions with schoolchildren as young as seven, and elderly communities in their eighties and nineties. I have run sessions on the boardroom tables of the world's largest retail businesses, and I have run sessions with homeless communities under road bridges, sheltering from the snow. Every object handling session I have run has unique aspects, of course, and no two are quite the same. The objects brought by participants are wide-ranging, but always "evocative," a term that Sherry Turkle, Professor of the Social Studies of Science and Technology at MIT, uses to describe "familiar objects, which become part of our inner life."[6] Possessions like these evoke powerful memories and meaningful associations within the minds of their owners. Through these sessions, I have seen thousands of different objects over the years, and what people will choose to bring is always unpredictable. For example, I recall one session where a designer for a global fashion brand brought a white T-shirt, which she purchased at a supermarket. She described how she used to like wearing these cheap T-shirts at home for doing the painting and decorating. One afternoon, while she was up a ladder, painting the ceiling, her boyfriend walked in and proposed to her, paintbrush in hand. The T-shirt now reminds her of that important life moment, but it also reassures her that she is enough, just as she is. Another object story that stands out in my mind came from a father who brought in a small blanket. Wrapped carefully inside the blanket was a model made of about a dozen multicolored Lego bricks. We were asked kindly not to touch the Lego model, but we were free to look. The small object sat cushioned on its soft blanket while he tearfully recounted the story of his young son's battle with leukemia, a battle the little boy tragically lost. Years later, the father gathered the strength to go through his son's old toys and give them to friends, family, and local charity shops. He told us how, in the center of the Lego box, he found this "work-in-progress" model. A precious object, indeed, one that connects him powerfully to his son, and to a time in his life when he was well enough to play. What starts out as a random collection of stuff—a cheap white T-shirt, a small cluster of Lego bricks—can soon transform into treasures of immeasurable value and significance.

Many participants bring objects that were lost and then found years later. One individual told the story of a ring given to her by a great-grandparent when she was nine. She misplaced the ring soon afterward, and despite turning the house upside down, she never found the missing item. Almost a decade later, when she was packing her things to leave home for the first time, the ring reappeared. The meaningful association she had with the object had intensified and strengthened significantly since the loss. She suspects that her great-grandmother's ring "hid" for a time to teach her to be more careful. Now reunited, she and the ring are inseparable. She describes how it supports her in trusting fate, bringing her back to a more mindful, cautious way of living. This story is not unique. Many people receive similar "life lessons" from the experience of losing and then recovering possessions. Their preciousness always rises sharply on being recovered, as the period of loss contributes a rich chapter in the object's unfolding story, and vital lessons are learned through it. Similarly, on a cultural level, "Some of the most unique and meaningful objects from history have survived not by intention, but by being lost and then found at an opportune moment. The Dead Sea Scrolls, the Rosetta Stone, and the Antikythera Device never would have made it to modern times without first being lost."[7]

Participants in object handling sessions themselves undergo powerful transformations in the way they perceive the various items that others brought along. At the beginning of the session, they see a random pile of worthless stuff, the likes of which we might expect to come across in a thrift store or garage sale. By the end of the session, after the sharing of stories, participants see a collection of priceless, irreplaceable artifacts to be treated with extreme care and respect. Indeed, although the types of objects people bring are varied, the reasons behind their significance are surprisingly consistent, even in radically different cultural contexts. Here I am reminded of Elizabeth Chin's extraordinary autoethnographic work *My Life with Things*.[8] Chin is a professor at Art Center College of Design in Pasadena, whose work engages marginalized youths in collaboratively taking on the complexities of the world around them. Her book offers a brutally honest account of what it means to share space with material things, and the entangled, complicated nature of the stories and meaningful associations we collaboratively form with our things.

Although the meaning of self differs cross-culturally and varies in its link with individualism,[9] that these conceptions of self are expressed through objects seems to be universal.[10] Indeed, despite the diversity of objects at these sessions, what is striking is that the stories behind each one share a common thread—a thread that unites every individual participant, from every part of the world. This thread is "connection." Our cherished possessions serve as conduits through which we experience connectedness with people, places, and times. One might say, things that matter to us connect us to things that matter to us. Objects provide experiential bridges connecting us to worlds we wish to maintain a relatedness to. The existence of these bridges becomes especially apparent when unwanted negative experiences flow across them. For instance, when an item of clothing connects us to the betrayal of an ex-lover, we will often discard or destroy it. When we dispose of such objects, we effectively demolish the bridge and thus remove the connection. Through discarding behaviors, we disconnect ourselves from associations we experience as negative: we cut off from them, creating distance. In a more positive way, these experiential bridges built by our cherished possessions maintain important connections with aspects of our lives that are beneficial and meaningful. These bridges, and the positive experiences that flow across them, are an essential component of our emotional well-being. These connections to people, places, and times are always interlinked, and we never experience one without the other. For example, receiving your grandmother's watch as an heirloom object connects you to her (people), but it also elicits childhood memories (times) of playing with the watch while watching TV in her apartment (places). We also see this interlinked character play out through souvenirs. One might assume that an Eiffel Tower snow globe connects you to Paris, but it equally connects to the time you went to Paris, the people you encountered along the way, and the things that happened to you while there. Occasionally the "person" in the story is you, and you alone. In this case, the cherished object connects you to a former self that you wish to remain connected to, and a memory you wish to remain present in your life. And so we live "with" things. A small handful of these things might be cherished possessions with high levels of inherent value. The majority, however, live as anonymous products kept largely for their instrumental value. These majority objects are never truly taken possession of and, as such, remain on the periphery of our experiential worlds. Out there on the experiential margins, products are all too easily discarded. When they break or require maintenance, we simply are not motivated to invest in them. And when newer, faster models come out, it is all too tempting to upgrade to a newer, younger model.

The experiential borderlands separating products from possessions are seldom crossed. The holy grail of marketers and product developers, these borderlands remain revered but poorly understood. Indeed, the process of "taking possession" is one of the most fundamental processes in shaping our relationships with the designed world. Without exploring this process, we can only superficially understand the purpose of selling, gifting, and even stealing objects,[11] or what it means to experience remorse when a lost possession is gone from your life. It is not enough to dismiss material things as mere products, outputs of some commercial system or a unit of exchange between global producer and local consumer. This framing presents us with an unhelpfully narrow range of possibilities when it comes to understanding what it means to share physical and emotional space with designed things.

Sometimes referred to as "psychological ownership theory," individuals may enter a mental state wherein they feel an object is truly "theirs."[12] Occasionally we even see such objects as a part of ourselves, and it is in this state that products become possessions—a part of what Russell Belk refers to as an "extended self."[13] Belk is professor of marketing at the Schulich School of Business. His work explores the meanings of possessions, and how ideas about ourselves are both shaped by, and represented in, what we desire, own, and discard. What most people overlook is that this moment of "possession" does not occur at the point of purchase. Nor does it end at the point of disposal. Rather, the mental state of possessiveness is fluid and can be attained for material or immaterial objects; it can even be experienced in relation to things that do not legally belong to you.[14] Individuals may, for example, feel a sense of ownership for things they do not own,[15] like the Tesla they intend to buy but have not yet raised funds for. Conversely, we may never take possession of things that are legally ours, like an unwanted birthday gift we feel obliged to keep.[16] We might also keep things that we no longer feel an affinity to. And so, on an emotional level, we have dispossessed them, even though we keep the object and use it from time to time. Just as products can be promoted to the higher status of "possessions," so too can possessions be demoted back to mere "products." Indeed, any object that is still in a person's legal ownership but has been emotionally dispossessed is only waiting temporarily before it is properly disposed.[17] Although the object does not change because of its passage from product to possession, our meaningful associations with it are entirely transformed.

Cultural Norms in Keeping and Discarding

We live in a throwaway society. But who is the "we" in this statement, and which society does it refer to, anyway? Can we really generalize about all people in this way—all 7.8 billion of us? Are we saying, for example, that the four hundred million people of the seventeen countries across the Middle East, with their nuanced and rich cultural, political, and religious histories, all share identical materialistic patterns and tendencies? Or that the three thousand different ethnic groups, speaking more than 2,100 different languages, throughout the fifty-four countries of the African continent—a continent home to some of the world's poorest countries, like Malawi, Burundi, and South Sudan—all live in a wasteful, throwaway culture? Of course one cannot say this. Yet this is not a "developed world" versus "developing world" dichotomy, either. Even within a capitalist nation like the United States, we see huge discrepancies in living conditions, which in turn influence the character of a person's relationships with objects in terms of how "throwaway" things are. So, no, we do not all live in a throwaway society. I do, and it is possible that you do too, but not everyone does. Indeed, aggregating all people as a "we" is about as pointless as describing all people as mammals. While it provides a consistent macro level on which we are all connected, beyond that, it is useless. This paradox of people being related and unrelated at the same time reminds me of a beautiful line by the American philosopher and psychologist William James: we are like islands in the sea, separate on the surface, but connected in the deep.

But it is not the inaccuracy of the statement that concerns me here. Rather, what is most troubling is the somewhat colonial attitude from which the statement originates. Too often, overgeneralizations like this place an unhelpfully "universal" (singular) distinction atop the world. This hegemonic, monocultural worldview narrows the range of cultural possibilities and assumes hierarchical dominance over all others. The depth of colonial assumption behind statements like this is problematic. It assumes a universality, a singular cultural worldview, in which all people share the same reality and inhabit the same world of excess and wastefulness. Indeed, while many societies do have a tendency toward wastefulness, the majority do not. In earlier writings—and in countless lectures too, no doubt—I have used this problematic phrase, repeatedly. Back then, in my younger years, it never really occurred to me that, from my position of white, male, northern European privilege, I was casting aspersions on the world as though it were one, singular thing, and speaking as though my reality was the only one that mattered. This kind of sweeping, generalizing rhetoric is both grossly inaccurate and deeply unhelpful.

The idea of culture has always been complex; today it is becoming increasingly so. With runaway globalization, catalyzed by omnipresent social media, many consumers are becoming members of multiple sociocultural milieus and hold multiple social identities, which in turn influence their judgments and decisions.[18] In an increasingly connected world, the idea of "culture" is becoming increasingly diffuse. Many of us lead double lives and are simultaneously global and local. Our previously isolated, singular cultural identities are becoming ever more entangled, hybrid, and unclear. Culture is complex. It is about far more than what country or countries your parents come from, where you were born, or where you live today. Culture is more helpfully understood as the shared characteristics and knowledge of a group of people, which encompass a great many things, including language, religion, cuisine, social habits, music, and arts, for example. It is the shared pattern of behaviors and interactions, cognitive constructs, and understanding that we learn through the process of socialization.[19] Thus cultural formation is established over time, through ongoing social interactions, often mediated through objects and environments, within a group. It is through these group interactions and social and material practices that we design, make, and transform the world around us. And so "no matter what culture a people are a part of, one thing is for certain, it will change."[20] Objects play a central role in this emergent, continual process of cultural formation.

Many social scientists view culture as consisting primarily of the symbolic, ideational, and intangible aspects of human societies, of which material things play a dominant role. Yet the essence of a culture lies not in its artifacts, tools, or other tangible cultural elements but in how the members of the group interpret, use, and perceive these things.[21] In other words, what defines culture is not the things themselves but the way we interpret those things. It is the values, symbols, interpretations, and perspectives that distinguish one people from another in modernized societies; it is not material objects and other tangible aspects of human societies.[22] What is more, people within a culture usually interpret the meaning of symbols, artifacts, and behaviors in similar ways.[23] For example, we commonly find the same products in radically different cultural settings. Although the tangible object is the same, the way in which the object is understood can vary dramatically, depending on the cultural context through which it is encountered. Of course, culture is not fixed; it is fluid and constantly in motion, albeit a slow and incremental form of change. Many cultures are adapting toward increasingly wasteful lifestyles, through rising affluence and an increasing level of access to products and services. In the boardrooms of the corporate world, these shifting cultures are cynically referred to as "emerging markets." Indeed, the commercial pursuit of greater knowledge and understanding in relation to cultural diversity is seldom driven by a desire to offer users more culturally sensitive experiences or to expand the sociocultural richness of one's material encounters. Rather, commercial considerations of cultural nuance are generally motivated by a desire to penetrate untapped markets—primarily in the global South—and create new products that resonate deeply with an unfamiliar customer base with money to spend. Such is the way capitalists capitalize on those with capital.

Into each discursive engagement between subject and object, users introduce their own unique cocktail of preconceptions, beliefs, and ideals. Through our interactions with them, objects reflect the subtle differences of each individual. For example, a cockroach will elicit different meaningful associations from a hygiene inspector than it will from a reptile enthusiast deep into the practice of herpetoculture. The "object" is the same, but the affective response, and the meaningful associations that form, is anything but. In this way, assigning meaning to objects is flexible, since it does not derive from the physical characteristics of the object. Like "dialectical variation" in language, the same object will have different meanings to different people because of its different associations to them.[24]

It is well understood that our past shapes our present. Individually, each prior experience performs on its own to guide and shape future action. We learn from experience by using the past to inform the future. Collectively, however, these prior individual experiences coalesce to form an aggregate package greater than any individual item. Together, this aggregate package shapes our worldview, values, and mind-set. It situates us experientially and shapes the very nature of engagement between people and things. In this way, product experience, and our experience of the designed world, is both individually and culturally determined. Indeed, it is already well understood that different people respond differently to a given product. This is because people differ with respect to their concerns, motives, abilities, preferences, and goals.[25] It follows, therefore, that their affective response to a given product will be similarly erratic. Experience is thus not a property of the product but the outcome of human-product interaction, and therefore dependent on what temporal and dispositional characteristics the user brings to the interaction.[26] That is to say, experience does not lurk within the object; rather, affective responses are activated within us through our interactions with designed things—a form of cultural process in which our interactions with the world stack up over time to form behavioral schemata that shape our values, choices, and actions.

Although numerous research studies show a correlation between culture and designed experience, the precise nature of this relationship remains unclear. Like experience, culture is a complex and layered construct.[27] Key determinants of culture—such as "values"—play a central role in the formation of product experience. The novelist and critic Raymond Williams, for example, describes a socially mediated form of culture in which culture is simply a description of a way of life, which expresses meanings and values, not only in art and learning but also in institutions and ordinary behavior.[28] Williams's extensive writings on politics, culture, media, and literature helped establish the field of cultural studies. He believed we might begin to understand a culture if we look closely at the explicit and implicit meanings and values emerging through our interactions with the designed world. These interactions with the world tap into our intrinsic and extrinsic values, which themselves have been shaped through the cultural conditions we inhabit and have inhabited. They are diverse and continuous and range across designed and nondesigned worlds. These interactions contribute to establishing "cultural norms," which may be understood as the shared expectations and rules that guide the behavior of people within social groups. Cultural norms, in this sense, are learned from and reinforced by parents, friends, teachers, and others as we grow up in a society.[29]

An object's instrumental value is not particularly influenced by cultural norms. For example, a knife is sharp or blunt whatever culture it finds itself in. In contrast, the intrinsic values elicited by an object are deeply culturally determined. As Marx defined it, the instrumental value of a thing resides in its utility—its ability to satisfy a human need or want. Intrinsic value, on the other hand, is typically considered to be a latent, rather than concrete, dimension of an object that is determined by its ability to generate and signify meanings within specific social or cultural contexts.[30] This relates to the French anthropologist Pierre Bourdieu's theory of "cultural capital," in that consumption is, in part, an expression of cultural expertise.[31] Bourdieu was particularly interested in how power dynamics are embodied in society, and how these systems of power last across generations. He proposed the central idea that cultural capital functions as a social relation within an economy of practices and comprises all the material and symbolic goods that society considers rare and worth seeking.[32] In his 1986 essay "The Forms of Capital," Bourdieu describes three forms of cultural capital: "embodied cultural capital," comprising the nontransferable knowledge acquired through socialization to culture and tradition, which shapes character and way of thinking; "objectified cultural capital," comprising possessions that symbolically convey a person's wealth, in addition to the acquisition of other forms of cultural capital; and "institutionalized cultural capital," comprising formal recognition of cultural capital, such as academic qualifications or achievements, providing a heuristic with which to evaluate a person's sociocultural status.[33] The term "capital" is useful here, as it suggests that one might accumulate a "wealth" of knowledge and skills, just as one might amass other forms of wealth, such as money. It follows, therefore, that individuals with different cultural experience and background will seek out different forms of symbolic value in the products and services they have access to. Research in the sociology of culture, and in the anthropology of consumption, suggests that goods and services carry a symbolic value to the extent that they provide users with a means to express individual identity and to signal social status to others.[34] Just as the determinants of social status are culturally variable—different from place to place—so too are the objects and material practices deployed to mediate this status. We see this play out in the way different cultures value education as a form of cultural capital—something valuable to have and something worth striving for. Cultural practices like studying and teaching are highly respected and valued in this kind of cultural environment, owing to the dominance of the cultural norms that hold "an education" in such high regard. In contrast, a cultural setting in which education holds lower levels of cultural capital will see less striving toward this goal. In turn, the professions that support education will be held in lower regard, as they fall outside the system of value of that precise cultural setting. In this way, culturally sanctioned goals can be seen to set the direction for societies to pursue.

That commercial products acquire a wide range of sociocultural meanings has been long recognized and studied by researchers in sociology, anthropology, and cultural studies. Consumer behaviorists, for example, have shown that these meanings influence how an individual might value and consume products. Consumers have been found to purchase and use a variety of products ranging from cultural goods to consumer electronics to express their social identity, defined as actual or desired membership in sociocultural groups. Perception of product value therefore depends on the congruence between a product's cultural meanings and an individual's social identity.[35] Individually and culturally meaningful relationships with the designed world can enhance well-being and enhance our experience of everyday life. Melanie Wallendorf and Eric Arnould describe the range of ways in which material objects play such important roles in social and cultural life. Objectively, products fulfill a broad range of multifaceted roles. They serve as tools to accomplish tasks. They provide mobility. They counterbalance the effects of nature by keeping us dry when nature is wet, warm when it is cold, cool when it is hot, shaded when it is too sunny, and in the light when it is too dark.[36] Subjectively, on the other hand, products serve as the set and props on the theatrical stage of our lives; they provide markers to denote our characters for others; they remind ourselves of who we are while communicating these ideas to those around us. In this sense, we derive our self-concept from objects, as they both "mirror" and "project" glimpses of the self, to ourselves and to others.[37] On the subjective level, objects perform in this way, veiling an underlying flow of social relationships and connections to places and times that matter.[38] Meaningful objects serve as "personal storehouses of value and significance."[39] Yet, as the professors of psychology and sociology Mihaly Csikszentmihalyi and Eugene Rochberg-Halton argue, when civilization becomes increasingly mediated through our associations with designed things, we find ourselves in psychologically precarious territory. They argue that things are cherished not just because of the material comfort they provide but also for the information they convey about the owner and the owner's connections to others.[40]

Of course, a degree of standardization may be necessary in fostering more fluid connections between geographically and culturally diffuse peoples and possessions. Certain digital tools, like Facebook and Twitter, for example, elevate users' ability to share their world and experience. Yet as material culture migrates toward an increasingly singular, mass-produced ideal, we see a requisite move away from locally made things; the adoption of culturally alien objects is an inevitable, if regrettable, part of this acculturation process.[41]

In *Making Things Better*, the anthropologist David Napier demonstrates how anthropological description of non-Western exchange practices can offer a tonic for contemporary economic systems in which our impersonal relationship to "things" transforms the animate elements of social life into inanimate sets of commodities. According to Napier, our impersonal relations to things—and to other people—have become so ingrained in our being that we take them for granted as we sleepwalk through a mass-produced and overly standardized life.[42] Indeed, we must ensure that design, and design education, is not further homogenized into a singular dominant paradigm, an authoritarian and universal system in which all must fall in line with the dominant Anglo-European model of reality, at the expense of the richness and diversity of all other cultural life. Arturo Escobar—discussed at the outset of the book—similarly calls for a radical reorientation of design, from its universally functionalist and commercially driven applications within globalized capitalist societies toward a culturally expanded view of design that is more in tune with the radical interdependence of all life. He argues for a postpatriarchal and postcapitalist "pluriverse"—a world where many worlds fit, contrary to the current model of a single, market-driven globalized civilization.[43] His inspiring treatise calls for a more open and inclusive mode of design and designing. Reinstating such a radical, abundant design culture, in which we reclaim design for other world-making purposes, requires creating a new, effective awareness of design's cultural embeddedness.[44] Like a growing number of others,

Escobar calls for an entirely different way of life, and a whole new style of world making. He highlights the urgent need to transition design culture away from the closed monism of neoliberal globalism, toward a more open form of design founded on principles of a significantly expanded pluriversal imagination. This will not be easy. As Escobar himself says, "It is easier to imagine the end of the world than the end of modernity."[45] In line with his view that all good design offers an alternative to how things currently are, Escobar knows all too well the scale of the task he lays out—not just its far-reaching global implications, but even more so its requirement for a fundamental shift in how we think, imagine, and understand the experience of being in the world. Or, as Escobar asks, what would it entail to construct a non-Eurocentric design imagination?

Everyday Acts of Product Maintenance

An important aspect of keeping—of living "with"—things is maintenance. Keeping things requires a degree of ongoing and periodic care to ensure possessions remain operational. This may range from light maintenance acts, like polishing a tabletop or installing a new operating system, to intensive maintenance acts, like stripping and cleaning a carburetor or expanding the RAM module on an aging laptop. Even when products receive frequent care and attention, they still occasionally fail. Such breakdowns require us to repair them, and if we cannot repair them ourselves, we must find someone to fix them for us. There are two types of maintenance: "preventative maintenance," which occurs before a failure has occurred; and "corrective maintenance," which occurs after a failure has occurred. This distinction is important, as notions of maintenance and repair are often incorrectly assigned as oppositional ideas. Rather, maintenance and repair are aspects of the same thing, occupying different ends of a "spectrum of care," with preventative maintenance (e.g., washing and ironing a shirt) at one end, and corrective maintenance (e.g., replacing a broken button) at the other.

As discussed earlier, the norms and values underpinning our relationships with the designed world are culturally situated and differ significantly from one person to the next. Through the lens of product maintenance, these cultural distinctions become particularly apparent. Whether we are talking about the kind of maintenance that supports the day-to-day upkeep of a product, or the kind of maintenance that returns a product to a prior working state, the practice of product "care" is socioculturally situated. We see different trends in product care practices around the world, not just in where it happens most or least, but also in the discrete ways in which these care practices are perceived by the individuals doing it, and the people around them. In much of the global North (the United States, Canada, Europe, and developed parts of Asia, as well as Australia and New Zealand), the active process of corrective maintenance (repair) has all but disappeared from everyday life. Many users lack the tools, space, and skills to engage in the repair process, and for others, the will to maintain even the simplest of appliances, garments, or items of furniture simply is not there as they become socially conditioned out of the making process. This situation is worsened by the fact that modern products are not designed to be easily repaired and are physically impenetrable to most users. Yet we can also understand maintenance as a state of mind, one we are continuously engaged in, whether watering a wilted houseplant, sharpening a blunt kitchen knife, cutting your hair, or messaging a friend you have not spoken with in while. Maintenance, in this way, is about taking care, making small efforts to keep things good and look after them.

People often wrongly assume that maintenance and repair practices are all but dead in our modern, streamlined world. While this may be the case in many domestic product ownership scenarios—like trashing a torn bedsheet rather than sewing it back together, or tossing a dining chair because of its wobbly leg—the practice of maintenance and repair is alive and well in the technical infrastructure required to enable services. Think large, complex infrastructural systems like the energy grid, public highways, or the national rail network. All these "products" are undergoing continual maintenance, upgrade, and repair. These renewal practices are so normal we barely notice them. When it comes to consumer products, however, it is as if we forget all of this and revert to disposability and material short-termism.

We have all faced the dilemma of repair versus replacement. It can be tempting to throw something out rather than repair it. For many, product repair is a frustrating and demoralizing process, which often ends in failure, further damage, or a sense of having wasted time. Occasionally users welcome product failure because of the way dysfunctionality provides a socially acceptable justification for prematurely discarding an unwanted object, for instance, a poorly chosen gift you previously felt obliged to keep but are now set free from because it is "broken."

Placing repair at the center of the design process enables a powerful shift, whereby we significantly increase the life span of products simply by making them repairable. However, products are often designed to be difficult to fix. Electronics have become harder to repair, partly because they have become more complex as technological assemblages. In the main, though, their "closed-off" character is by design. Manufacturers have increasingly restricted the availability of repair information to authorized repair centers, leaving consumers and independent repair services unable to deal with even the simplest of technical problems. December 2018 saw the first ever "right to repair" protest as EU member states gathered in Brussels to vote on proposals that would oblige manufacturers to make home appliances easily repairable and longer lasting by design.[46] We are seeing more pushback from users as concerned citizens gather together to demand their right to repair. The use of "repair prevention" as a method of making products obsolete is growing.[47] Manufacturers have several reasons for this. In most cases, it is cheaper in the short term to manufacture a product that is not designed to be disassembled—its parts repurposed, and its materials recycled. The assembly line will have fewer stages if something can be more crudely assembled using nonreversible snap fittings or glues, as opposed to screws, which occupy space and are comparatively time-consuming to install. Even when screws are used to hold parts of a product chassis together, whether a power drill or a flat-screen TV, they have idiosyncratic tamper-resistant heads, meaning you could not unscrew them even if you wanted to. For those screws that can be removed, they are often self-tapping and therefore so aggressively threaded that, once removed, the receiving threads are stripped to the extent that the product can no longer be pieced back together.

From the mid-1900s onward, products became increasingly specialized. With this, they also became harder to repair by anyone other than a skilled artisan or specialized craftsperson. This brings us to the importance of the community; repair flourishes within networks of knowledge, materials, and skills.[48] The rise of "repair cafés" and "fixer communities" around the world is a testament to fact that we need a network of skills and support if our highly specialized products stand any chance of being effectively maintained and repaired. Today, with increasing levels of automation in the supply chain, products become ever more inaccessible to the layperson. As a guiding principle, products assembled by humans are easier to dismantle, maintain, and fix than products assembled by machines. Once inside, few people would know what to do with the complex assemblage of componentry anyway. The "usability" principles applied to most electronic products only apply to their slick exterior. If you do get beyond the exterior shell, product interiors are unintelligible no-go zones that simply were not designed for you: private domains to be inhabited only by assembly line workers, in sterile environs, at the point of manufacture. Increasingly, crossing the threshold from the exterior to the interior of a product renders warranties invalid and, as such, severs any ties with the manufacturer.

A cynical motivation lurks behind the closed nature of modern products. After all, if they cannot be repaired, we have no choice but to replace them with new ones. Naturally, this stimulates further consumption, drives sales, and prevents anyone from keeping something for too long. In January 2019, Apple CEO Tim Cook announced in a letter to shareholders that people are buying fewer iPhones because they are repairing the ones they already have. This, according to Cook, is a key reason behind Apple's failure to meet sales targets set for the final quarter of the 2018 fiscal year. Citizen-led repair was not factored into Apple's product planning or service design process, and so these grassroots activities went on, but without Apple's participation. Apple was effectively cut out of the process because of its apparent lack of interest in product repair. Consequently the company missed significant value creation opportunities.

In most cases, products are not designed to be disassembled. We see this at recycling facilities, where truckloads of unwanted appliances and orphaned electronics are smashed to smithereens using heavy industrial equipment—chains, grinders, crushing rollers, and wrecking balls—that seems more suited to architectural demolition or quarry work than to dismantling hi-tech electrical equipment. These imprecise and somewhat "medieval" mechanical processes occur at the end of a product's life and provide a sharp contrast to the way such things were painstakingly assembled. On the assembly line, products are pieced together in silent, dust-free clean rooms by latex-gloved workers and state-of-the-art automatons. The two opposing ends of this story could not be more different. With coveted precision at one end, and barbaric destruction at the other, this tale has an unsettling Freudian dimension to it, exposing the point in the supply chain where we believe value to be (creation), and subsequently where we believe value no longer exists (destruction). Once products are demolished, the result is a low-grade aggregate of metals, glass, polymers, and a host of other materials, all impossibly fused together at the time of their making. Various processes, from magnetic sifting to vast floatation tanks, attempt to separate and extract the more economically valuable material, but really, most of these materials were inseparably bound together at the point of their manufacture. In most cases, they end up being downcycled into low-grade postindustrial applications like road surfacing or carpet tiles—hardly a fitting destination for some of the world's most advanced, high-performance materials.

Making an Effort

A greater sense of ownership develops when users participate in the process of making, maintaining, and repairing an object. Maintenance itself is a practice, and a direct correlation exists between the practice of maintenance and how attached someone is to the object in question. At the level of making, consumers develop a deeper sense of connection to objects they have played some part in producing. This phenomenon plays out in the most basic of assembly tasks, like inserting the ink cartridge into the shaft of a new fountain pen or threading the laces into a new pair of shoes. User involvement in the process of creation therefore deepens subsequent relationships with products. This process is often referred to as the "IKEA effect." This phrase, coined by the professors of business, administration, and behavioral economics Michael Norton, Daniel Mochon, and Dan Ariely, denotes a cognitive bias in which consumers place a disproportionately high value on products they have partially created or fixed.[49] These three authors noticed that effort alone can be enough to induce greater liking for the fruits of one's labor: even constructing a standardized bureau, an arduous, solitary task, can lead people to overvalue their often poorly constructed creations.[50] Similarly, Leon Festinger, a social psychologist best known for his influential work on cognitive dissonance and social comparison theory, claimed that the more effort someone puts into something, the more that person will value it.[51] Similarly, when we receive a gift that has been handmade, we value it more highly. The receiver of the gift vicariously experiences the effort, skill, and time the giver invested in the object's creation, and as a result, the recipient values it more highly. But this is about more than just effort; it is also about competency. For example, the process of building a self-assembly dining chair helps us to feel competent. Through the completed chair, we might then display that competency to ourselves and others. We see this phenomenon in the case of home cooking. For example, when instant cake mixes first came out in the 1950s, homemakers resisted the new product because the instant mixes made cooking "too easy," which in turn made their effort and competency as a cook feel undervalued. Because homemakers did not feel sufficiently invested in the baking process, they put no value on the product. In response to this problem, the producers of the cake mixes made a simple change in the recipe: homemakers were required to add an egg. By adding one more step—cracking an egg—homemakers felt as if they were now actually baking, which resulted in increased sales of instant cake mixes.[52]

Product maintenance supports design that lasts in objective and subjective ways. Objectively, maintenance keeps things in good working order, and through this process, possessions tend to have a longer service life. Subjectively, the restorative act of product maintenance rekindles our affections for objects while providing a feeling of competency that is demonstrated through the object, both to oneself and to others.[53] These objective and subjective factors work together to postpone our eventual compulsion to replace older things with newer things.

Users can interpret the failure of object maintenance as the inadvertent "sabotage" of an object,[54] like a form of well-intentioned vandalism, through which an attempt to make things better has accidentally made them far worse. We then must live with the consequences, as the object provides living proof of the user's incompetency and poor judgment. The object becomes a source of shame—a reminder of the moment we tried to fix something and failed, memorialized in the object. In this way, failed maintenance experiences are disastrous for the longevity of an object. Such missteps in the subject–object relationship are ordinarily terminal and lead to the "ridding" of the object. One thing is clear: user engagement through maintenance enables a form of cocreation and therefore changes the way we experience and perceive objects. If we are to design objects and experiences that invite users to more actively participate in the process of maintaining them, it is vital that these experiences must be "easy wins," maintenance tasks that people can comfortably and easily perform. If not, then we are increasing the hazardous nature of object relations by incorporating yet another reason for people to become disenchanted with their possessions.

In their paper "Practices of Object Maintenance and Repair," the social geographers Nicky Gregson, Alan Metcalfe, and Louise Crewe recount the story of an elderly couple's TV cabinet, damaged by builders doing renovations to the couple's home. A small flake of hardwood veneer was accidentally chipped away from the cabinet's surface, exposing the low-cost chipboard substrate beneath. The builders denied having done this, causing great anger and frustration for the owners. The flake of veneer was missing, and they did not feel they would be able to restore the object to its former state. In the end, the TV cabinet had to be replaced with an identical new cabinet. Importantly, the replacement was not motivated by a need to maintain a pristine house, or through any dislike of less-than-immaculate possessions. Rather, the replacement was fueled by the couple's need to distance themselves from the negative experience and unwanted meaningful associations the damaged cabinet continued to elicit, even months after the fact. Interestingly, the owners believed the cabinet to be insufficiently damaged to be declared "rubbish," and so they gave it to a neighbor,[55] who would of course not have those negative meaningful associations with the object. On the contrary, the neighbor probably had positive feelings toward it owing to the warm, neighborly circumstances through which he or she acquired it.

Repair need not always be a regressive process of restoring something to a former state. Oftentimes, repair is a progressive process, pushing something forward into a coproduced and expanded form. This kind of "transformative repair" is repair that changes an object's appearance, function, perception, or signification.[56] Kintsugi provides a helpful example. Kintsugi, which translates as "golden joinery," is the centuries-old Japanese art of fixing broken pottery with a special lacquer dusted with powdered gold, silver, or platinum. It treats breakage as a natural part of the history of an object "being in the world," rather than something to hide or disguise—a progressive, rather than restorative, process. Beautiful seams of gold glint in the cracks of ceramic ware, giving a unique appearance to the piece.[57] This Japanese ceramic repair practice includes the potential capacity of repaired objects to transform an audience or public within a cultural context.[58] While the process is associated with Japanese artisans, the technique was also applied to ceramic pieces of other countries, including China, Vietnam, and Korea. We also see ceramic mending practices elsewhere, beyond Asia. In eastern Iceland, for example, a traditional form of mending cracked stoneware bowls uses undyed thread. Broken fragments of ceramic are hand-stitched back together, each "cross-stitch" acting as a staple, clamping the broken pieces together like sutures to a wound.

Traditional repair methods, such as darning and patching, focus on making repairs that are both visible and relatively easy to perform. They require low levels of skill to carry out but yield high levels of objective and subjective reward. That is, these maintenance practices objectively repair the fault (it is not broken anymore) while subjectively expanding our experience of the object and its story (its history has been rewritten). This form of visible mending sees product failure as an opportunity to enhance the look and value of objects. Be it a knitted cardigan with frayed cuffs or a smashed porcelain vase, we can think of repair as a bold and partisan act of transformation, rather than a polite and regressive act of restoration. We also see these kinds of visible mending practices in fashion with the process of repairing holes, stains, or worn areas of garments in such a way that these "scars of age" are enhanced and even valorized. These acts can be quite simple, like replacing a broken white button with a new crimson one or darning a hole in a black sock with bright yellow thread. In these cases, the fix becomes a personal tribute to heavy use, a handmade declaration of love for worn objects. This is scar tissue we want, scar tissue that testifies to a moment in time when we chose to act in a way that declared independence from the aesthetic shackles of others. We stood out from the crowd and declared who we truly are. Autobiographical traces like these are potent, but unfortunately few and far between. This is because we have been designed out of our material worlds and banned from participation. As users, we are relegated to the role of passive button pusher, an obedient spectator. All we can do is pay for the show, sit back and gawp in awe at the brilliance of others, meanwhile reflecting on our own failings and inadequacies as individuals.

Clearly, complex social, environmental, personal, and economic factors draw people toward—or away from—maintaining possessions. One thing is for sure: care for material things is waning. Although repair thrives in socioeconomic situations of material poverty, it flounders in so-called advanced economies where it is cheaper and easier to replace broken products with new ones,[59] particularly when industry continually taunts us into doing so.

This represents a huge cultural shift. Throughout the twentieth century, firms aggressively promoted planned obsolescence and designed things to break. As the economist and critic of American consumer capitalism Victor Lebow wrote in 1955: "Our enormously productive economy demands that we make consumption our way of life, that we convert the buying and use of goods into rituals, that we seek our spiritual satisfaction, our ego satisfaction, in consumption. We need things consumed, burned up, replaced, and discarded at an ever-accelerating rate."[60] Indeed, the act of repair is about far more than thrift or eco-efficiency. It can reorient how we think about our own values, and how our values relate to the values of other people around us.[61] Indeed, the challenge is not technological or even economic but social and cultural. Can we make it delightful to maintain and preserve things?[62] Importantly, this is not a question of duty or of "guilting" people into repair. Rather, it is about reconnecting with the partisan act of maintenance and repair to activate more experientially rich modes of engagement with the designed world, and each other.

3   Material Matters

The Periodic Table in Your Pocket

The hidden story of materials runs far deeper than what the eye and hand can perceive. Peer beneath a product's glossy, scratch-free skin, and a theater of resources is performed at an atomic level, with unfathomable depths of ecological and social consequence. Within the last century and a half, economically aggressive corporations have mined, logged, trawled, drilled, scorched, leveled, and poisoned the earth, to the point of total collapse. As Robert Macfarlane eloquently describes in *Underland*: "Our modern species-history is one of remorselessly accelerated extraction, accompanied by compensatory small acts of preservation and elegiac songs. We have now drilled some 30 million miles of tunnel and borehole in our hunt for resources, truly riddling our planet into a hollow Earth."[1] Macfarlane is a British scholar of English who writes at the intersections of landscape, nature, place, and memory. He weaves a crosscutting angst—and at times rage—throughout his work, which unites these large, complex contexts with a unifying concern for the anthropocentric desolation of the world.

In manufacturing consumer electronics—arguably one of the most socially and ecologically significant areas of design activity today—we produce forty tons of waste to manufacture one ton of products. Of that ton of products, 98 percent are discarded within just six months of purchase. In the use of energy and material alone, this sequence of events is less than 1 percent efficient. Or to put it another way, our prevailing system of production and consumption is over 99 percent inefficient. This is completely unacceptable and begins to explain why, in an ecological sense, we are in the perilous situation we are in. This does not just apply to high-tech products like smart thermostats and laptop computers; even low-tech products like toasters, hair dryers, and electric toothbrushes are hideously complex when understood from the perspective of their constituent materials.

All electronic products are the culmination of international labor from mines, refinery facilities, and assembly lines, labor usually provided by underpaid workers. Hundreds of thousands of people work in dozens of countries around the world to extract resources and refine the materials used to make modern electronics.[2] Everyday electronic products have complex designs, also containing a relatively large circuit board, stuffed full of rare earths locked permanently in position. "There are 17 rare earth elements, which are embedded in smart products like tablets and smartwatches, making them smaller and lighter. They play a role in colour displays, loudspeakers, camera lenses, GPS systems, rechargeable batteries, hard drives, and many other components."[3] From yttrium and samarium, to gadolinium and ytterbium, valuable rare earth elements are woven throughout even the most whimsical electronic device.

A cheap radio-controlled child's tank, for example, contains a thumbnail-sized microchip within which you will find over two-thirds of the elements of the periodic table. Let us not forget, these elements do not come out of the ground in clear glass test tubes, labeled and ready for use. On the contrary, these rare compounds are mined and clawed from the earth, encased in thousands of tons of rock and mud. A hazardous, polluting, and energy-intensive process of extraction and purification follows, within which lurks the majority of our material world's ecological and social destruction. An industry in itself occupies this level of resource conversion, behind the scenes, transforming the earth's rare resources into miraculous modern materials. The plethora of electronic products that surround us are infused with a rich cocktail of complex minerals, precious metals, and noxious compounds. To most of us, these are completely unseen, lurking just beneath the surface of our material experience. Despite their genius, these wonders of industrial alchemy become waste surprisingly quickly, and they certainly are not designed to last.

Kate Crawford and Vladan Joler's extensive essay "Anatomy of an AI System" beautifully articulates the diffuse material nature of the products that mediate artificial intelligence.[4] Crawford and Joler's research roams the borderlands of technology, society, and environment. Using the example of Amazon's Echo, they dismantle the illusion of its immaterial, ghostlike presence of AI to expose the ecological and social burden AI systems like

Echo leave in their wake. They speak of the correlations between the mining of material and human resources. They point to the "deep interconnections between the literal hollowing out of the materials of the earth and biosphere, and the data capture and monetization of human practices of communication and sociality in AI."[5] They caution that while consumers become accustomed to a "small hardware device in their living rooms, or an app, or a semi-autonomous car, the real work is being done within machine learning systems that are generally remote from the user and utterly invisible to them."[6] The Echo, Crawford and Joler argue, "is but an 'ear' in the home: a disembodied listening agent that never shows its deep connections to the neural networks within Amazon's infrastructural stack."[7] Its physical presence is deceptively light, allowing the product to misleadingly command wonder in the immaterial, frictionless world it supposedly inhabits. This lightness is a deception. "Each object in the extended network of an AI system, from network routers to batteries to microphones, is built using elements that required billions of years to be produced. Looking from the perspective of deep time, we are extracting Earth's history to serve a split second of technological time, to build devices than are often designed to be used for no more than a few years."[8]

The majority of today's postconsumer waste is disposed of via community recycling centers and landfill sites, which leach liquefied heavy metals and other toxic contaminants over time, such as arsenic, cadmium, copper, lead, manganese, and zinc. These toxic compounds find their way into soil and groundwater, disrupting living systems and posing a significant threat to biodiversity. Landfills are also known to produce large volumes of methane (CH_4), a greenhouse gas and main contributor to climate change. In fact, CH_4 is twenty-five times more potent per kilo than CO_2.[9] A policy briefing by the Woods Hole Research Center in 2015 concluded that the Intergovernmental Panel on Climate Change (IPCC) does not adequately account for "permafrost carbon feedback." As the Arctic warms, the permafrost melts, soil decomposition accelerates, and the upper layers of this frozen soil begin to thaw, allowing deposited organic material to decompose and produce large quantities of CH_4. A smaller percentage of waste is burned using vast incinerators that produce ash, laden with toxic elements, while also releasing hazardous acidic gases such as sulfur dioxide and nitrous oxides into the atmosphere. Even the much-acclaimed recycling of discarded electronics consumes large quantities of energy, and the chemicals involved during treatment and sorting often find their way into living systems.

Precious elements like gold are commonplace in electronic products, largely owing to its excellent conductive capabilities. Alarmingly, there is approximately eighty times more gold in a ton of smartphones than there is in a ton of rocks from a gold mine. However, because of the way we have designed our industrial systems, the rock-bound gold is currently considered to be more economically viable to extract than its phone-bound counterpart. This kind of waste is a design flaw. With rising resource costs and mounting levels of legislation-driven producer responsibility, situations such as these must change. One might assume that, given the huge quantities of precious resources that find their way into our gadgets, we would take a little more care of them, fix them when they break, and keep them for longer periods of time. However, as we all know, this simply is not the case, as products simply are not designed with longevity in mind.

Electronics find their way into an astounding array of common objects that you might not normally associate with electronics: pens, bags, trainers, Christmas sweaters, keys, baby bottles, Lego, and even talking cookie jars. On my son's tenth birthday, he received a card from a friend. When he opened the card, it began to play "Happy Birthday" to him. Though the song brought a smile to his face, the card was promptly closed and placed on the pile of other cards before even reaching the end of the song. A short life, indeed, for a product containing just over half the earth's known elements. Yet this type of brief encounter is typical of our modern world. In fact, it is so typical that most people do not even notice.

Minerals and Conflict

With the exception of surface-based renewables, most materials start out as a giant hole in the ground, often in some remote, underregulated part of the world. Yet, unbeknownst to many, these resources can link us to violence, conflict, and suffering in many of these far-off places. In the linear economy (take, make, dispose), this social disaster unfolds at the beginning and end of product life cycles, but both ends engage in different forms of resource extraction practices. The first is concerned with the cheap extraction of virgin resources, and the second with the recovery of valuable materials from scrap. Indeed, the electronics in our pockets, bags, and homes connect us to destructive industrial practices via invisible threads of commerce, science, politics, and power.[10]

Concern for the environmental and social impacts of electronic products is now widespread.[11] Despite this growing concern, designers are largely ignorant of the ways in which their sourcing practices affect people across product supply chains. Even self-declared "sustainable designers" remain unaware that the mining of ore to produce tin, tungsten, tantalum, and gold fuels war and conflict in countries like the Democratic Republic of the Congo (DRC)—a country that in 2015 ranked second from bottom in the UN Human Development Index, making it one of the least developed nations on Earth.[12] While the mining sector is often known for social and environmental abuses, these four minerals have been singled out as being especially problematic. The DRC is rich in tungsten, tin, and gold, and the southern area of Katanga produces 80 percent of the world's cobalt, a mineral widely used in rechargeable lithium ion batteries. Overall, the country is estimated to have $24 trillion worth of minerals under its soil. Despite such enormous resource wealth, the country remains one of the most destitute places in the world, with one in seven children dying before the age of five, according to UNICEF.[13] Through a combination of deep government corruption, lack of regulation, and a highly exploitative mining industry, none of this wealth finds its way back to the people who inhabit and work these lands.

Throughout history, resources have often played a role in conflict; drugs, oil, and diamonds have all been connected with fueling conflict-related activities.[14] The term "conflict minerals" refers to raw materials that come from a particular part of the world where conflict is occurring and affects the mining and trading of those materials.[15] The sale of conflict resources enables a self-financing war economy based on mineral and human exploitation. Materials typically contained in a laptop, for example, originally enter the supply chain from the mining sector. Be it pollution, dangerous working conditions, or child labor, mining-related practices desperately require improvement in many parts of the world. Armed groups engaged in mining operations in this region subject adult and child workers to serious human rights abuses and use proceeds from the sale of conflict minerals to finance regional conflicts.

The four most commonly mined conflict minerals are cassiterite (for tin), wolframite (for tungsten), coltan (for tantalum), and gold ore. Collectively, these minerals are referred to as 3TG. Of course, conflict resources are widespread and span beyond the scope of 3TG, for example, blood diamonds from war-ridden Angola, Ivory Coast, and Sierra Leone; conflict timber from Cambodia; or ISIL terrorist activities funded by oil revenue in the Middle East. Even water is being described as the conflict resource of the future, and though a wide range of water-related conflicts have appeared throughout history, rarely are traditional wars waged over water alone.[16] Indeed, 3TG minerals are so widespread in the manufacture of consumer electronics that they demand much closer consideration by design. It seems irresponsible to be in the business of creating objects without at least some sense of the human stories behind the resources you use.

So what roles do these precious 3TG minerals play in our electronics? First, tin is a widespread material used in a range of applications from food packaging to transportation. Its dominant use in consumer electronics is in solder for making electrical and mechanical connections between components on integrated circuit boards (ICBs). Second, tungsten is incredibly durable and hard-wearing, used in chip manufacture in smart watches and a host of other digital products caught up in our thrust toward an ever more quantified self. Tungsten is a heavy mineral, which makes it an effective counterweight in small vibrating motors. It is also an extremely hard, dense material and has the highest melting point of all metals. Third, tantalum is rare. Often used as a substitute for platinum, tantalum is an inert, corrosion-resistant, and highly stable element. In electronics, it is used primarily in capacitors, semiconductors, and ICBs. It is also dosed into alloys to increase stability and resilience. Finally, gold is one of our most highly valued materials. Not only is it loaded with perceived cultural value, but it is also corrosion resistant, extremely malleable, and highly conductive. Its uses in electronics are as bond wire between ICBs and as plating for connectors. Collectively, these properties make 3TG minerals highly sought after. Even the largest global electronics manufacturers depend on a steady flow of these "miracle minerals."

Multinational firms such as Intel and Motorola have unwittingly purchased conflict minerals in the past to manufacture complex digital products.[17] Confusion prevails as to the best approach to responsible sourcing. Indeed, 3TG presents a highly complex legislative context where attempts to stem illegal mining activities often do more harm than good. One example of this is the Dodd-Frank Act (2010) enforcing transparency and accountability in companies to ensure their products are free of 3TG from countries involved in resource-funded conflict. When the act was proposed, its supporters said it would weaken the militias by cutting off their mining profits. However, once passed, the legislation triggered an unforeseen chain of events that has since propelled millions of artisanal miners and their families deeper into poverty by removing their only source of income generation.[18] It is widely believed that a shift to products that are 100 percent conflict free can only be achieved by adopting a strategy of sourcing all materials from outside the DRC. The flip side of this boycott option is that many regions in the DRC that are completely uninvolved in conflict will be adversely affected by this decision. In 2016, 70 percent of tin, tungsten, and tantalum mines in the eastern Congo have "conflict-free" status.[19] Even in the mines affected by conflict, this approach would simply eliminate one of the only economic opportunities available to local people.

Of course, it would be possible to eliminate or significantly reduce the use of these materials in the design and production of most electronic products. Fairly simple design-led changes can be made, such as using copper bond wire in place of gold, or omitting tungsten altogether by dropping the vibrate function. While it might be technically possible to design out conflict minerals, it is hard to see how this would actually improve the lot of the people directly affected by the conflict.[20] However bad we might imagine life to be for an artisanal miner, removing people's only livelihood option leads to widespread school dropouts as people can no longer afford fees; malnourishment spreads, and even rudimentary health care becomes unaffordable.[21] Indeed, this is a truly wicked problem.

Many businesses fear greater levels of transparency in supply chains and would rather not talk about it. Unethical corporate practices like these rely on low levels of consumer awareness and understanding for such exploitative profiteering to freely continue. An industry-wide transition toward positive changes in materials supply chains is under way—a transition in which we seek to better understand the issues, source more responsibly, increase use of recycled materials, and actively seek partners who can help achieve these goals. As more businesses begin to recognize that their supply chain is also a designed object—albeit a distributed one—a shift toward better practices is beginning to emerge. Clearly, engaging with customers in transparent and open dialogue about the people, culture, and politics inextricably woven throughout our possessions is necessary. Those who do follow best practices see this as an opportunity to celebrate and educate customers.

Industry leaders are opening up supply chains and initiating discussions about where their products come from, how they are made, and by whom. Growing public awareness of the gruesome side effects of the trade in conflict minerals has nudged some of the big tech companies to reallocate their investments away from the most suspect big mining concerns, investing instead in new generations of small artisanal mines that are relatively "clean" in terms of militia links, corruption, and working conditions.[22] Take a common household product like an electric kettle. Laden with conflict minerals, most kettles are not designed to last. They are "closed devices," difficult to repair. If even the smallest of components breaks, users hastily assume the entire product needs to be trashed and replaced. This is not the consumer's fault; it is just how we have been educated to engage with electronic products. This is one of the normalizing features of our dominant sociotechnical paradigm, whereby conspicuous consumption rages unchecked, enabling the blind pursuit of further technological innovation.

Waste Pickers of the Global South

Millions of people worldwide scratch out a living by collecting, sorting, recycling, and selling materials from the mountains of unwanted products we discard. Waste picking provides a source of income to many desperate, marginalized people—street children, orphans, the elderly, widows, migrants, the disabled, the unemployed, and victims of armed conflicts. Waste picking also benefits the broader economy by supplying raw materials to industry and creating many associated jobs for middlemen who purchase, sort, process, and resell materials collected by so-called waste pickers.[23] There is growing recognition that waste pickers contribute to the local economy, to public health and safety, and to environmental sustainability. In this respect, one can see some benefits to all of this, particularly in recycling and the recovery and resale of high-value materials. However, waste pickers ordinarily face deplorable living and working conditions.

Many sociocultural complexities characterize this kind of work and the way livelihoods are interwoven with the waste. Many urban centers of the Global South—from Belo Horizonte in Brazil to Accra in Ghana— would simply not function were the rubbish pickers not there. That being said, industries must not automatically advocate such practices simply because they reduce the burden of city councils and national legislators in relation to waste management. Separating materials that have been soldered onto circuit boards is costly work, and this complex form of e-waste—though responsibly sent for recycling by the user—ends up getting shipped to an unregulated country with little or no environmental law. After reaching these lawless shores, our waste is picked over by the fingers of children who painstakingly salvage traces of semiprecious material from the carcasses of our unwanted gadgets. Once stripped, these carcasses are stockpiled, buried, or burned. Entire families live atop these smoldering heaps of electronic waste, which spew clouds of noxious fumes, so there really is no minimum or maximum age for working as a waste picker. These makeshift furnaces are not used for cooking or keeping warm. Rather, the burning serves a purpose for the waste picker by providing a primitive recycling technique, which smelts away the plastic to uncover the valuable metals within, exposing workers and their families to deadly substances. A trend toward specifying plastic parts instead of metals in electrical and electronic goods is also causing problems for the health of waste pickers because so many of these plastics use toxic flame retardants.[24] Brominated flame-retarding chemicals have been associated with lower mental, psychomotor, and IQ development, poorer attention spans, and decreases in memory and neural processing speed. Chemicals such as these are commonplace in modern consumer products. One study found significant traces of two potentially hormone-altering brominated flame retardants in 43 percent of children's toys, including toy robots, hockey sticks, and finger skateboards.[25] Such substances are even more commonly found in polymers used in electronic products, which are less tightly regulated than children's toys and games.

Today several international initiatives are addressing global e-waste management and trade concerns, as well as issues with environmental pollution caused by e-waste. Recently introduced EU law now prohibits the export of e-waste, but not the export of refurbished materials. As we know, even refurbished smartphones reach end-of-life at some point, so the waste still ends up in these places eventually. There are exceptions to this rule. Some products are responsibly dismantled, disassembled, and recycled via well-considered product take-back schemes operated by manufacturers. So why is it important to talk about these issues—surely they fall outside the scope of design concern? The situation is a conflated one, shaped by a complex array of previously disconnected issues and practices. Primarily it is a consequence of an ineffective global policy environment, our wasteful behaviors as consumers, and a systemic failure in design to consider end-of-life. It is as though we avoid talking about death, preferring to avoid the subject altogether. In describing "closure experiences," Joe Macleod, author of *Ends* and self-confessed "endineer" (one who engineers better endings for things), illustrates how consumers and providers tend to overlook the importance of healthy, coherent product endings. We once had a rich culture of reflection and responsibility, but over recent centuries this has been lost, producing a mixture of long-term societal oversight and short-term denial, says Macleod. Endings are dodged and left for someone else to clean up.[26] While products occupy our hearts, minds, and pockets, they perform the roles for which they were created, and we love them for that—for a while, at least. However, the world of goods casts a dark shadow. These underworlds have operated flagrantly for decades but remain outside the awareness of most designers. The majority remain entirely ignorant of these inhumane and socially destructive practices and would likely be horrified if they knew. At least I like to think they would.

The designed world provides a gateway into savage depths of human suffering. During the all-too-fleeting use phase, it is easy to overlook all of this. In affluent societies, certainly in northern Europe, it is easy to forget that the recycling bin is not the end of a product's journey but, more accurately, a new beginning. Most of us would like to imagine that our unwanted products are rounded up in silent electric trucks, sorted into neat piles, and dismantled and recycled by happy and healthy workers in clean, well-lit industrial environs. Maybe the radio is even playing in your imaginary recycling facility? I joke, but this is only because to most of us, these places exist purely as imaginaries, constructed with incomplete shards of information we assemble over time. Most of us have never, and will never, see one of these places. Places like this exist, but they are rare, even in the handful of countries with strict waste management policies and environmental regulatory systems. In increasingly affluent societies like Brazil, India, and China, the situation could not be more different.

When it comes to discussions about products that last, we often hear of ambitious design intentions, played out through relatively unambitious product typologies. In this scenario, radical visions for design that lasts (e.g., modular design, graceful aging, repairability) are undersold through technologically oversimplistic product types (e.g., self-assembly shoes, denim iPad cases, pimped-out bike repair kits). These are valid design responses and often show how theoretical ideas might work in practice, but they do not adequately point to where the locus of our collective design attention must be. In particular, they sorely overlook electronic products altogether—arguably the most problematic product type of all. One notable exception to this norm is Fairphone, the world's first ethical, modular smartphone.

Fairphone is a Dutch social enterprise that works toward the design of conflict-free modular smartphones. It is a well-used exemplar of sustainable product design in European design schools, yet relatively unknown elsewhere in the world. Fairphone sources conflict-free tin and tantalum from mines in the Democratic Republic of the Congo and works with manufacturers to improve working conditions in its factories. The company also sources responsible gold from Uganda. Fairphone believes that "to contribute to more responsible gold mining in Uganda, we must address a variety of interconnected factors. For example, we aim to tackle the issues at the root of the child labor—ranging from access to education to better incomes for adults. At the mines themselves, we want to increase economic prosperity with training and better mining equipment."[27] Products like this turn the tide in a sector hell-bent on specifying the cheapest materials with little or no regard for the ethical dimensions of their provenance, then irreversibly gluing them together to save time and money. In addition to ethical sourcing practices, Fairphone also focuses on durability, repairability, and the availability of spare parts that can easily be replaced to extend the smartphone's usable life. In the case of a product that is to be leased and refurbished several times during its operational life, designers have the expertise to obtain intimate knowledge of how the product and its parts wear and tear, and how to decide what parts should last (slow parts), and what should be replaced (fast parts), and when (service touch point).

The notion of a product being renewed and replenished in this way is not as bizarre as it may initially seem. For example, as I think about Fairphone, I am reminded of a pair of thirteenth-century Shinto shrines near the city of Ise in Japan. Both shrines are rebuilt from scratch every twenty years according to an ancient Shinto tradition. This practice not only keeps the structures intact, even when made out of relatively ephemeral materials like wood, twine, and thatch, but allows the master temple builder to train the next generation.[28] The sixty-second rebuilding of the temples was completed in 2013, and the sixty-third rebuilding will take place in 2033: a slow, cyclical process that incorporates vital elements of ritual in an ongoing process of renewal, spanning generations. Parts of the "object" are slow and others fast, and these ritualized forms of ownership and care uphold the object's completeness. This harkens to the classic thought experiment in the metaphysics of identity called the ship of Theseus. Here the question is whether an object that has had each of its components replaced remains the same object. The story goes that the famous ship sailed by the hero Theseus is stored in the harbor as a museum object. It rots, and over time, parts are sequentially replaced. Eventually none of the original ship exists; is it now a replica, or is it still the original? The human body is like this, too. It might surprise you to know that the human body regenerates itself entirely once every ten years or so; organs take about a year, skin about two weeks. So are you the same person you were ten years ago? Well, the answer is both yes and no.

Fairphone's founder Bas van Abel believes that opening the design processes behind everyday products like smartphones will allow people to reengage with their possessions and become more active owners of their stuff. Such is Fairphone's commitment to this idea that the phone itself even says, "Yours to open, yours to keep," on the back casing. Formerly head of Waag Society's Open Design Lab, Van Abel sees design as a pathway to radical new forms of social and environmental entrepreneurship. Fairphone successfully marries the triple concern of ethically sourced materials, long-lasting design, and a more complex user relationship with the product and its story. The modular design is just complex enough for most people to be able to dismantle the phone and make small repairs. All components and their modules can be added and removed using a conventional Phillips head screwdriver. What is more, commonly failing components like batteries or displays can be replaced without any tools. If you design your phone to last twice as long, you only need to produce half as many phones. If your business model is based on selling large numbers of phones, this idea does not work. However, if the business model is based on leasing phones and services, longer life spans are highly beneficial. Fairphone describes how "putting ownership in the hands of the manufacturer gives more control over the device and puts Fairphone in a better position to take advantage of the circular economy. Because if Fairphone leases the phone instead of selling it, they can ensure that all the resources inside are used optimally over the course of the phone's life cycle, including when it is time to be used by another client or recycled."[29]

The Base of the Iceberg

Products are far bigger than they at first appear to be. Like icebergs, their size and scale far exceed what their surface characteristics might at first suggest. They are formed over time in ways that reflect their provenance and their origins. And, like icebergs, products have a life span, but because of the unpredictable conditions and circumstances they find themselves in, some last longer than others. Yet, beyond their majesty, icebergs are potentially deadly, too. There is the part you see, and the part you do not, with the submerged portion of ice representing nearly 90 percent of the total ice mass. As any mariner will tell you, the part of an iceberg you need to worry about is not the part you can see but the greater submerged portion—the part that will tear a catastrophic gash down the length of your ship's hull.

Picture such an iceberg, floating in the frigid sea. Now, assume you could extract all the water from this mental image. What you would be left with is, effectively, a large block of ice. No waterline, no buoyancy, no parts you can see or parts you cannot see—just an elephantine mass of ice. This solid lump is the whole iceberg, out of context, and imagined in a way we most likely have not experienced before. Now let us think about a simple product like a pair of denim jeans. These jeans float effortlessly through our everyday lives, revealing to us only a very small fraction of their true ecological and social burden. At the exposed surface of this object, we might notice some of its more blatantly obvious features: textile quality, pattern cutting, particular shade of blue, for example. However, many more of this product's impacts lurk beneath the surface, situated just beyond our field of awareness. You probably won't notice the 1,500 gallons of water that were required to grow enough cotton to make just one pair of jeans; that the rivers in the Mexican textile-manufacturing city of Tehuacán now flow a bright toxic indigo, largely devoid of life; or how your very own laundry habits are responsible for about 50 percent of the environmental damage caused by this pair of jeans throughout their life span. When experienced in their totality, products are far greater than we initially perceive them to be. From the user's perspective, perhaps this is not such a bad thing. After all, we would be unable to function if our awareness was so finely attuned to the nuance of every material transaction of our lives. A simple task like opening a can of beans would take several hours, owing to the crippling cognitive load placed on us when reflecting on the human consequence of the morally questionable supply chain behind the ink on the can's crudely printed label. However, from the designer's perspective, this "blindness" represents a deeply problematic rupture in our somewhat narrow understandings of the work and its wider social and environmental implications.

Attention works in this way. Like shining a spotlight in a darkened room, it reveals only small parts of a far wider space: as the beam pans left and right to illuminate another area, it simultaneously plunges the previous area back into darkness. Similarly, as we sit on the bus going home, to fill an idle thirty minutes, we reach for our smartphone. As we tap the Facebook icon, there are a few things we are consciously paying attention to, and a far greater number of things we are not. As your Timeline unfolds before you, you might reflect on how little social attention you seem to be getting, or how unpopular your post from last night was, which you really thought was going to have a far greater impact among your network of so-called friends. You might feel pangs of inadequacy, or guilt, on witnessing the freshly baked tray of brownies held aloft by an old friend's five-year-old daughter. You decide to do more fun things like that with your kid . . . just as soon as this big project ends. And so your mind is awash with thoughts and ideas triggered by even superficial encounters with physical and digital artifacts. But these are the things you are actively thinking about. What you are not thinking about is that about fifteen thousand kilometers overhead is a satellite, no bigger than a Mini Cooper, hurtling through the thermosphere, relaying signals from the palm of your hand down to an energy-hungry server farm on another continent, ensuring that this steady flow of anxiety-inducing media flows seamlessly through your electronic device. You are also not thinking about the electromagnetic waves this product pushes through your body or the endocrine-disruptive particles emanating from the polymer film you bought to protect its fragile screen.

In *Phase Media: Space, Time and the Politics of Smart Objects*, James Ash argues that objects and interfaces create envelopes, or localized foldings of space-time, around which bodily and perceptual capacities are organized.[30] Ash's work focuses on the cultures, economies, and politics of digital technologies and their transformative sociocultural influence. This ensemble—or "envelope"—that Ash highlights is ordinarily geared toward the explicit production of economic profit, but the potential of these spatiotemporal encounters is greater than this. Although they focus our attention, these localized foldings of space and time also inadvertently create sizable blind spots in our mental models and represent deeply flawed and inadequate imaginaries of the material systems that shape our world. Jay Wright Forrester, computer engineer and pioneer of system dynamics modeling, described how mental models are "the image of the world around us, which we carry in our head. Nobody imagines all the world, government, or country. They have only selected concepts and relationships between them, and use those to represent the real system."[31]

As a continually unfolding process, "Design is a conversation about what to conserve and what to change, a conversation about what we value. It is a process of observing a situation as having some limitations, reflecting on how and why to improve that situation, and acting to improve it."[32] And yet, "When a designer tries to bend something to their will, a tension pushes back. We not only have a relationship with our materials; we are a material too. We are here, falling through time and taking up space. Breathing in air and consuming food. We cannot exist without taking."[33] Considering a more expanded spatiotemporal dimension of objects enables us to reconceive alternative imaginaries of what objects are, beyond their immediate material boundaries. Through this expanded frame, we might more effectively reflect on their ethical implications. Such imaginaries are the creative and symbolic dimension of the social world, the dimension through which human beings create their ways of living together and their ways of representing their collective life.[34] As socially assembled systems of meaning, imaginaries are fictional constructs, in that they are not necessarily provable or true. The impacts of these imaginaries on our values, behaviors, and choices are very real. Indeed, the ways in which people think about, understand, and imagine concepts such as denim jeans or Facebook, and the agency people perceive they have, affect what they do. Dan Lockton and Stuart Candy, design professors and foresight professionals with deep expertise in how we imagine, experience, and understand present and future worlds, argue that these imagined versions of reality shape our behaviors; these fictions are political, and they matter.[35] Similarly, the prominent anthropologist, cyberneticist, and systems thinker Gregory Bateson said that "the world partly becomes—comes to be—how it is imagined."[36] Design plays a central role in imagining the products, systems, and processes that constitute our material reality. This material world is an emergent property of our collective values, beliefs, and aspirations; it is our collective values made manifest. So if the world becomes how we imagine it, as Bateson says, then influencing how we imagine the world must become the focus of our endeavors. Bateson refers to this relational failure—between the world and our understanding of the world—as "epistemological error."[37] We find this sentiment also in the work of Alfred Korzybski, a Polish American scholar of semantics who famously stated that the "the map is not the territory," encapsulating his view that an abstraction derived from something, or a reaction to it, is not the thing itself.[38] Through more expansive and persuasive visions of the future, we might expand our cultural imaginaries, introducing new ways of understanding our worlds and the myriad designed artifacts within them.[39]

The Horse, Not the Phone

One of the greatest illusions of our time is that of the seemingly shrinking material world. As products get smaller, we might assume a requisite decrease in their environmental burden. This, unfortunately, is not the case. Take the cell phone, for example, a product that began its life in the 1970s resembling something more like carry-on luggage than the sleek, pocket-sized devices we have come to know and love. At the time, cell phones were barely portable, weighing almost a kilogram, costing several thousand dollars, and, in some cases, offering little more than twenty minutes of talk time.[40] Yet despite their cumbersome form, early cell phones had considerably smaller environmental impacts than even the simplest of handsets on the market today. Though these early phones did far less from a functionality perspective, they used far fewer resources than even the most basic of modern devices. They also had significantly shorter supply chains, with locally sourced materials and manufacturing processes being the norm.

Like a shadow, this illusion of dematerialization sits behind much of the stuff we own. A typical smartphone today, for example, weighs about 210 grams, give or take a few grams depending on the make and model. The product's weight is made up of a number of different materials and components, and most people would be able to guess only a handful of these. Some polymers are easily identifiable in the casing and screen—they contain alloys that help give the device rigidity and physical durability—and it also contains some complex circuitry within, which we will never see but can assume is there. Finally, we know it has a battery inside because we have charged it up. So if you weigh all of that, it comes out to around 210 grams. However, this is not actually the case. The 210-gram smartphone is an illusion, a trick. It actually weighs more like five hundred kilograms, or half a ton. That is the true weight of this product, when you consider all the materials flowing through it, and all the resources required to bring this object into the world. If we calculate the per-unit weight of the whole product story (e.g., resource extraction, material processing, component manufacture, product fabrication, shipping, distribution, retail, use, eventual disposal), it weighs about half a ton. Approximately seventy-eight kilograms of this weight reflect the CO_2 generated as resources make their way around the world, flowing through various processes and stages of the supply chain. Of course, numbers can be a little abstract, so to put this into some kind of context, the average horse weighs about five hundred kilograms. Think about that for a moment. Every time you lift your phone to your ear to take a call, you are effectively lifting a horse, albeit in a disembodied sense. So sustainable product design is not about the 210 gram smartphone; it is about the whole 500 kilograms. Or, to put it another way, design is about the horse, not the phone.

This is a problem, as in design we have become so focused on the object that we have lost touch with the wider systemic conditions from which the object originates. In *Rethinking Society from the Ground Up*, Randy Hayes, executive director at Foundation Earth, says that these unseen externalities (e.g., resource depletion, energy consumption, pollution generation, social injustice) must be internalized into the price you pay for goods and services; only then will the ecologically cleanest be the cheapest. Hayes argues that when we can do this, we will have more of a "true cost economy" that factors in the living and nonliving systems implicated in our production and consumption.[41]

As in the example of the iceberg, we tend not to see the whole product. Rather, our mental models form around the parts of the product we most fully identify with. If you ask someone to draw a tree, for example, he or she will most likely draw the part of the tree that stands above ground. What you get, essentially, is a drawing of half a tree. It would be unusual for someone to draw the whole tree, including the roots. These flawed mental models of the exterior world are commonplace today, presenting a skewed mediation of reality as may be objectively described, a form of "fragmentalism" in which the world is experienced as a set of separate parts.[42] In contrast to holistic interpretations of phenomena, this reductionist worldview sees us carving up the world into smaller and smaller pieces until we reach such a fine level of subdivision that we can no longer effectively understand how these isolated fragments relate to their wider systemic conditions.[43] Through this mechanistic framing of the world, we have developed a distorted perception of reality and lost our ability to understand the holistic, interconnected nature of complex global crises.

Meaningful Materials

Too often we hear interrelated notions of "meaning" and "material" clumsily smashed together via hastily derived observations, stripped of context or situated dependencies. This leads to the production of naive sets of universal design principles, which assume a rational and altogether predictable character in our experience of materials. Reductive generalizations abound. Take concrete, for example: a material typically consigned to impersonal notions of coldness, construction, and the built environment; a material ordinarily found on the ground or in walls. Yet this is more a symptom of the way in which we have encountered concrete to date than a future-oriented take on the potential of concrete to engender meaningful new associations with users. These kinds of limitations in material thinking are commonplace and represent a failure to capture the highly unstable character and full creative potential of our materials experience as situated in the world.

Of course, concrete is well known as being an ecologically destructive material, responsible for around 6 percent of all CO_2 produced worldwide, and about 10 percent of all industrial water use. Yet while the scale of these negative consequences is remarkable, we must understand them in relation to the immense volumes of concrete used. Concrete is the second most widely used material on the planet, next to water. And so design judgments about what constitutes an appropriate material, and what does not, must always be contextualized and considered carefully. I hear people rejecting concrete as a material for small-scale, long-lasting products based on its ecological credentials—such as those I just mentioned—and yet freely specifying 3TG rare earth elements without a second thought.

Prior experience is a significant part of how relationships with materials, and their requisite meanings, form. However, we must not take prior experience as any sort of limitation on what materials can and might do in the future, on an experiential level. For example, when concrete is specified on a different scale, in an alternative typology of object, and situated in a different place, alternative meanings suddenly emerge. A concrete pendant light, like those designed and made by the Romanian design studio Ubikubi, jettisons much of the associative baggage of its material to enable a new appreciation of the material's qualities—its air bubbles, its glassy surface, its crumbling edges, its weightless suspension in midair—that expand our experience of the material. Or consider a concrete pen, such as the one by the Taiwanese design firm 22 Studio, which has a coldness to the touch and a weight to it that brings your attention to the object as you write. As can be seen, material meanings are idiosyncratic and wholly subservient to their immediate context. When an object typology shifts, or any other disruption in contextual familiarity behaves in this way, we refer to "material incongruence." Such incongruence helps to disrupt assumptions and creates space for us to reconsider new interpretations of otherwise familiar situations.

Of course, many alternatives out there support designers in making better material choices. In the fashion industry, for example, we see a surge of interest in materials like mushroom leather such as Mylo, which can be grown in molds to create seamless forms, or protein-based yarns inspired by spider silk such as those used in the Adidas by Stella McCartney biodegradable tennis dress. These lower-impact choices provide biodegradable alternatives to the havoc wreaked on our planet in the name of style, from "chemical treatments and dyes that add to air-and-water pollution, to animals bred in order to be fashioned into bags or shoes."[44] We are also witnessing rapid growth in markets for responsibly sourced rare earth minerals. Salmon Gold, for instance, is a partnership launched in 2018 that produces gold while restoring habitats for salmon and other species in Alaska, the Yukon, and British Columbia. Sustainable sourcing programs of this kind are receiving significant investment by metals-intensive companies from jewelers like Tiffany & Co. to electronics giants like Microsoft.

Yet, beyond their technical capacities, materials possess a deeper and more subjective dimension. In this way, materials are simultaneously technical and experiential—mediating ideas, beliefs, and feelings while delivering performance, functionality, and utility. No simple principles exist to adequately explain the complex interrelationship between meaning and material. For instance, it is not possible to locate a design method that will guarantee material x will evoke meaning y in product z.[45] Meaning depends on context.

All materials have meaning. Or, more specifically, all materials interact with their context of use and, through this interaction, elicit meaningful associations within the mind of their user. The meaning of a given material is a relational property, involving interactions between users, designed objects, and the materials from which they are formed. This notion aligns with Ludwig Wittgenstein's belief that "the meaning of artifacts lies in the particularities of use,"[46] and therefore any theory of meaningful materials must reflect the wider role of context in the meaning-making process. Context, in this sense, refers to a "situational whole" from which we ground the meanings we attribute to our world.[47] In describing the dependencies of text and context, Klaus Krippendorff describes how most words are ambiguous by themselves. A professor of cybernetics, language, and culture at the University of Pennsylvania, Krippendorff notes the number of diverse meanings a dictionary typically lists for a single word. In the context of a larger discourse, however, word meanings are usually singular and clear. Equally, by themselves, artifacts may not mean much unless they are placed in a particular environment in which they play recognizable roles.[48] In this way, meaning is unstable and entirely at the mercy of contextual factors.

One thing is clear: experience is constituted via the dynamic and continually changing interrelationship between people and their environment. The interplay of these constituent parts represents the totality of a person acting, sensing, thinking, feeling, and making meaning in a given setting.[49] The American philosopher and psychologist John Dewey argues that meanings are constructed in our interactions with objects, and both an object's formal properties and the individual who perceives the object play a role in the construction of meaning. In this scenario, diverse types of experience come together to form what Dewey calls "an unanalysed totality."[50] Through this totality, no distinctions are drawn between the material and the action or the subject and the object. Rather, at the visceral level, the world is encountered as an aggregate, experiential whole. On later reflection, the unanalyzed totality might be untangled and understood as a collection of isolated, nameable strands, like a piece of tartan. On experiencing the color, pattern, and overall feel of the textile, we do not necessarily consider the independent roles of each individual thread. On reflection, we might later consider this level of detail, but initially, at the visceral level, we simply experience the tartan as a complete, unified whole. We might also consider the different "levels" at which we experience materials, and the ways in which these levels activate different forms of meaningful association within users. For example, the prominent researchers of "materials experience" Elisa Giaccardi and Elvin Karana claim that no matter what the resulting experience for different individuals is, all materials are experienced at four experiential levels: the "sensorial level" (e.g., touch, smell, sound), the *interpretive level* (e.g., feminine, modern, traditional), the "affective level" (e.g., desire, curiosity, disgust), and the "performative level" (e.g., behaviors, actions).[51] Giaccardi and Karana argue that when designers consider all four levels during the selection process, they can achieve greater control over the resultant materials experience and the meaningful associations that form with it.

The attribution of meaning is dynamic and continuous. As a result, our understanding of materials, and the emotional affect they elicit, changes through use and over time.[52] Individuals thus understand their artifacts and interact with them on their own terms and for their own reasons.[53] Broader sociocultural factors, or the type of environment a person grew up in, will also play a major role in the way materials experience is extrapolated. For example, children whose first experience of the world comes from mass-manufactured toys and games from Walmart may develop a different set of material values and sensibilities than individuals who grew to maturity surrounded by handmade objects of wood, stone, and metal. My point here is not to say that one is better than the other but to say they are different, and our material sensibilities are more nurture than nature. Our individual histories are instrumental in shaping the experiential dimensions of our interactions with materials. As described earlier, it would be futile to derive a taxonomy of material meanings, articulating the meaningful associations that users will experience in response to a given material. Tempting as it might be to construct such a list—and many researchers do—it would be nothing more than academic folly, entirely divorced from any relational context. It would be like telling someone you know what the taste of horseradish will make them think and feel: pastoral English landscape, muddy walks down country lanes, roast beef, tradition? Sure, some thematic generalizations can be made, and occasionally you might even get it right, but these are things we think we think, rather than things we actually think. As we all know, our personal experiences introduce a level of idiosyncrasy that renders these normative principles grossly inadequate. For me, the taste of horseradish evokes sharp childhood memories of awkward Sunday lunches and complicated family arguments; such is my experience of this peppery condiment.

4 Deeper Experiencing

The Depths and Shallows of Product Experience

Too often, "experience design" serves up a dull and servile rendition of material reality, in which most encounters with the designed world come with life's mysteries and challenges already solved. The field has become distracted by the notion of seamless interaction—a misplaced "holy grail" in which users use things without even realizing they are doing so. This leads to the formation of a bland experiential terrain characterized by seas of near-identical products, each one beautifully obedient but horribly dull. They are what protein shakes are to home-cooked meals. The range of experiential possibilities enabled by products born of this mind-set is "pathetically narrow" and does nothing to cultivate a more meaningful and enduring culture of material things.[1] Yes, products fulfill a certain palette of needs and carry out tasks in an effective way, but from an experiential point of view, they represent a fairly barren wasteland of possibility. Furthermore, and perhaps more alarming for the task of steering design toward longer-lasting products, this dominant design approach takes us away from the peculiarities and idiosyncrasies of people and toward the anemic precision of autonomous, anonymous things. Usability is important, of course, but it represents the most basic level of engagement for any designer: to simply make something usable. Design has far more to it than this. Many professionals in experience design even believe their job is to make things easy to use—tangible products, software, digitally mediated environments, and so on—to make the world and the things in it more usable, seamless, and automatic. This move toward fluidity of interaction sorely misses the point; people are in this not for the speed and convenience of it but for the experience of it. Just as the job of a novelist is not to make books quicker and easier to read, so the experience designer's work is about teasing out threads of ambiguity, nuance, and richness to expand our experience of everyday life.

One of the mythologies underpinning this drive toward seamless interaction is that if only we could remove the noise and distraction of our clumsy and ineffective material encounters, we would be able to connect more fully with our inner selves. If only we could reduce the cognitive load placed on us by the many products that surround us, we might free up personal capacity to connect more fully with ourselves. This point is well intentioned, but a sorely misguided false opposition. Deep and complex encounters with material things provide a vital means of connecting with ourselves, of noticing ourselves, of experiencing ourselves, as mirrored through encounters with products. Just as interacting with people in ever more complex ways reveals just as much about us as it does about them, so material encounters also have a mirroring effect and provide an important existential function in that regard.

Over time, the design profession began to take the term "experience" more literally and wonder about what it would mean if we really designed for experiences as the primary output, rather than a nice flourish that can be added toward the end of the design process. But what is experience, and where can we find it? Primarily, experience does not reside in the object; it resides in us. Objects are only capable of shaping the nature and character of our experience or leading our experience toward certain outcomes. This may sound like an obvious thing to say, but it is an important distinction to make if we are to look closely at the design of deeper experiences. This common misplacement of experience has fairly serious consequences. Believing that experience somehow resides within the material lures us into focusing on the object even more and focusing on people even less. Through this deeper dive into objects, we become increasingly detached from the underlying principles and processes of experiencing, as human phenomena, and become sorely distracted by the props that shape these experiences. Indeed, experience lurks within us, awaiting activation by a trigger, a cue, or a prompt of some sort. Our external world provides such prompts but should never be confused with experience itself.

This focus on the material object can lead to highly abstract product propositions that fail to deliver from an experiential point of view. Swaths of mass-manufactured goods wash in and out of our lives, consistently failing to penetrate on any deeper level, completely inadequate in their affective capabilities and subtleties. In psychology, "affect" generally refers to all types of subjective experiences that are valenced, that is, experiences that involve a perceived "goodness" or "badness." In experimental research, valence is traditionally a bipolar dimension to describe and differentiate between affective states.[2] The nature of these affective experiences is shaped by the particularities of each individual user. One person might find owning secondhand clothing to be a positive and rewarding experience, while another may feel uncomfortable, and even repulsed, at the thought of wearing clothes used by another person. Of course, out in the world, products are experienced in a holistic and interlaced way. To the user, product experience unfolds continuously and largely without such isolated reflection.

To further understand product experience, we must first artificially disentangle it. Paul Hekkert and Pieter Desmet, professors at the Delft University of Technology leading research on design and emotion, break product experience into three fundamental levels: "aesthetics," "meaning," and "emotion."[3] They claim that the "aesthetic level" describes a product's capacity to delight one or more of our sensory modalities with how it looks, smells, sounds, feels, or tastes. We most commonly hear of aesthetics referring to how things look, but other forms of aesthetic experience also exist, such as a beautiful piece of music. The phrase "aesthetics of interaction" refers to this form of multisensorial beauty that one might experience when interacting with the designed world.[4] Hekkert and Desmet then propose that the "meaning level" is where cognitive processes like interpretation, memory retrieval, and association take place. At this level, they say, we can recognize metaphors, assign personality or other expressive characteristics, and assess the personal or symbolic significance of products. Finally, the "emotion level" is where affective phenomena emerge such as love and disgust, fear and desire, pride, and despair, to name a few.[5] These affects (emotions) powerfully shape patterns of consumption and waste by creating closeness and belonging, or distance and alienation, between people and their possessions.

There is nothing fuzzy or nondescript about emotions; they have a clear behavioral utility. Emotions pull us toward certain people, objects, actions, and ideas while pushing us away from others.[6] They help us position ourselves within the world of goods and support us in making course-correcting judgments that steer us toward products and lifestyles that will be better for us, and away from those that will be worse. In this way, emotions themselves are elicited by reflective evaluations (appraisals) of events or situations as being potentially beneficial or harmful.[7] These evaluations are not consciously carried out but rather occur just beneath the level of awareness. Thus it is our interpretation of the object, and not the object itself, that dictates the character of affective response. Objects merely provide cues or triggers to that affect; ergo, it is not the thing itself but the repertoire of prior experiences the object encounter conjures within you. An emotion, therefore, is the result of a cognitive, though often automatic and unconscious, process.[8] Furthermore, this process is highly subjective, shaped by each user's bank of prior experiences and memories. Different people can have very different emotional responses to the same thing.

Each of the three levels (aesthetics, meaning, emotion) provides an alternative lens through which sets of characteristics of experience may be more closely observed and understood. This separation exposes the foundations on which experience is constructed, and enhances the capabilities of designers looking to intensify the experiential capabilities of their work. Yet although each of these levels may be broken apart, named, and defined as individual elements, they do not actually function in this way. The lived experience is much messier and far more entangled in itself. The three threads of experience are simultaneously encountered and felt as an intertwined whole. One does not drive a car, for example, and objectively reflect on each experiential level one by one; the whole experience is far more entangled. In this way, we may understand product experience as an aggregate set of affects elicited through the interaction between a user and a product. As Hekkert argues, together these linked concepts offer significant value for shaping a more profound exploration of the complex, idiosyncratic, and rich experiences people have while interacting with products.[9]

Although product experiences can be unique, we also find archetypal stories that all of us can relate to, which connect these seemingly unique experiences. By studying these archetypes, we begin to shape an alternative grammar of product experience, which is simultaneously visceral and immediate (fast), and reflective and emergent (slow). The German psychologist Marc Hassenzahl studies the affective and motivational dimensions of interactive technologies. He tells us that experience-rich products have the power to transcend their encasing, as they contribute not to the aesthetics of "things" but to the aesthetics of "experiences." This is the challenge designers of interactive products face: user experience is not about good industrial design, multitouch, or fancy interfaces. It is about transcending the material. It is about designing and creating an experience "through" a device.[10]

At its best, experience design looks beyond the isolated individual product experience as a thing in itself, to engage more holistically with the complex experiencing processes of stories and memory. As the psychiatrist and literary scholar Iain McGilchrist proposes in his seminal book *The Master and His Emissary*, "Experience, like fire, is not a thing, it is a process, an unfolding."[11] What is particular to these unfolding stories is how my stories differ from your stories, how they are the same and how they are different, and how this helps me know myself and you. We even have our own memory system for archiving these stories of our lives, called "episodic memory": a person's unique memory of a specific event, as different from someone else's recollection of the same event. In kansei engineering (also known as affective engineering), the phenomenon of an interaction triggering an instance from episodic memory is defined as products being designed to "bring forward" an intended feeling from within the user. Experience design then becomes the curation of a series of design-led prompts that support the recall of prior experiences and the stories we associate with them. Though based on real-life events, these stories are largely fictional. That is, they are based loosely on what some may describe as an objective account of factual events and occurrences, but are massively distorted by subjective experience. In this way, our episodic memories are imagined versions of the event; accounts of a life lived, veiwed from each person's unique perspective. Kourken Michaelian, a philosopher of memory, explores the subjective, unstable nature of human remembering. He situates episodic memory as a form of "mental time travel," claiming that there is no intrinsic difference between remembering and imagining. He argues that "to remember is to imagine the past."[12] Michaelian goes on to explain how episodic memories shape not only how we imagine the past but also how we image the future—a phenomenon known as "episodic foresight." Episodic memory, therefore, is the crucible in which much of our imaging is formed, regardless of its forward or backward temporal orientation.

In noticing the experiential banality of material engagements, we must take care not to overreact. In *Enchanted Objects*, David Rose cautions against designing overly demanding human-object interactions. Rich material encounters "should be subtle, rather than insistent," he argues. Rose explores the ways in which digital information interfaces with the physical environment—for better or worse. To him, "Objects should never beep, buzz, or alarm. Instead, they should respect your attention like a polite butler cleaning his throat to get your attention."[13] Hassenzahl exemplifies this form of designerly discretion through the example of the Philips Wake-up Light—a bedside light that subtly simulates sunrise and sunset. Though not the best product in terms of environmental credentials, it does present a useful example of how richer forms of interaction design require carefully articulated stories and meaningful associations. We all hate our alarm clock, says Hassenzahl. When compared to the gently simulated sunrise and soothing birdsong of the Wake-up Light, the shrill tone of a conventional alarm clock feels a bit like having a bucket of ice water thrown over you every morning. The Wake-up Light fulfills the underlying need (waking up at a specified time) but does so in a way that "imports" experientially favorable episodic memories (being woken slowly by the morning sun). With this product, you set the wake-up time, and a half hour before then, the light begins to glow brighter and brighter. Then, at the moment you want to wake, it plays birdsong. The object thus connects us to a way of waking up that people feel good about, the memory of that one sunny morning on vacation when you were woken by the birds.[14] The product delivers a surrogate experience. As users, we know the product is merely simulating the experience of the sunrise and the birds, but we like the associations it creates within us. As Hassenzahl tells us, this surrogate is deeper and more real than an alarm clock ring, which is a purely technical and shallow solution that can only be associated with itself.

Hadal or Epipelagic?

The made world is a consequence—an emergent space in which the human species has progressively found ways to modify and enhance the world around us. The urban spaces we roam, buildings we inhabit, products we use, and garments we wear collectively represent our intellectual capacity to imagine a better world that is beyond our current level of experience. The innate capability to imagine a world just beyond our current level of experience and then formulate (design) plans to realize those imaginings is an essential determinant of what it is to be human—to reach beyond innate human limitations.[15]

We are each connected to several distinct systems of objects, occupying different depths of material experience. Possessions may well sit side by side on shelves, but they are divided by fathoms in an experiential sense. In searching for clues as to why some products last and others do not, we are drawn toward the things that matter: tattered old photographs of our children, handwritten notes from loved ones, the analog watch of a long-passed grandparent or the dress shoes you wore at your wedding. Like archaeologists, design researchers examine these emotionally durable artifacts in search of clues to their survival. The aim of this search, of course, is to discover the secrets of their success and uncover the designable factors that led to their transition from anonymous product to cherished possession. The assumption is that we might then apply these "design principles" to new products so that they themselves last longer. But how much can we really glean from these highly valued, idiosyncratic items, and to what extent are we establishing unrealistic expectations by placing them center stage in this way?

Using the metaphor of the ocean, with its distinct zones and depths, we might reimagine products in terms of the depths and shallows of the experiences they enable. A dive into this experiential world takes us through the swarming, dynamic assemblage of products occupying the surface, or epipelagic, zone, down to the select handful of precious possessions occupying the hadal zone of our deeper material world. Before we go any farther, and for this analogy to work, a little oceanography: In oceanic terms, the epipelagic zone comprises the uppermost 100 to 200 meters of the ocean, receiving enough sunlight for photosynthesis to take place. The epipelagic zone is far more abundant in marine life than the lower zones. In fact, it contains more life than all other oceanic zones combined. We know a great deal about the rich abundance of the epipelagic zone, as much of our research takes place there. In sharp contrast, the hadal zone extends from 6,000 meters to the bottom of the deepest parts of the known ocean, the Mariana Trench, at 10,911 meters. The water temperature in this forbidding zone is just above freezing, no sunlight penetrates its depths, and the pressure holds steady at eight tons per square inch. As with outer space, we know little about the inhospitable hadal zone, and only highly specialized life-forms, such as invertebrates like starfish and tube worms, survive down here.

Reframing product experience as having depths and shallows begins to expose previously obscured features that powerfully influence the character of relationships we form with products. Through our ongoing interactions with designed things, we experientially dive downward, from the dynamic, crowded shallows of the epipelagic zone to the quiet, sparse depths of the hadal zone. As we gradually dive downward through layers of experience and meaning, on-trend fonts make way for handwritten notes; digital screens make way for torn notepaper; and logotypes are elbowed aside by images of forgotten friends, lost family, and former lovers. Down here, objects become increasingly personal and idiosyncratic. It is sparse down here; everything moves more slowly and seems more stable. In contrast to the turbulent epipelagic zone, the enduring objects found in the hadal zone anchor us within a more constant, enduring notion of self. As a result, we cling to them dearly.

Designing longer-lasting products without a clear understanding of the depth at which you are engaging users is like packing a suitcase for a trip to an unknown destination and not knowing what you will be doing when you get there. The picture is incomplete and forces designers to make gross generalizations. As stated earlier, clear discrepancies exist between the products that matter, and the products we deploy as signifiers of identity and social position, for example. When we conflate these two aspects of material experience, results risk distortion, insights become misleading, and the impact of our attempts to create design that lasts remains limited. The debate surrounding the design of longer-lasting products has been hampered by a preoccupation with hadal items, lurking at the very depths of human experience. Through this bias, we overlook the weaker signals emitted by the myriad objects in the epipelagic shallows. These common objects characterize our experience of the everyday and fill the rooms, cupboards, and pockets of our daily lives. Arguably, they are also the objects that generate ecological and social pressure. In the depths and shallows of material experience, the level at which materialism manifests— arguably the site of human-made ecological destruction—occurs within the dynamic shallows. This epipelagic zone is populated by the plethora of mass-produced objects that fill our worlds. This constantly shifting assemblage of products is deployed to reflect our equally dynamic and unstable identities. As our identities evolve and change, so too must the products we deploy to both mirror and project these ephemeral ideas. Like a shadow that follows you around, this stuff defines you, whether you like it or not.

This form of materialistic value orientation involves the belief that it is important to continually pursue the culturally sanctioned goals of attaining financial success, having nice possessions, having the right image, and having high status.[16] In this way, materialism is defined as the importance a consumer attaches to worldly possessions,[17] or to a set of centrally held beliefs about the importance of possessions in one's life,[18] particularly their ability to communicate one's societal position to others. Most products occupy the shallows of our material worlds, as opposed to the more intimate mode of engagement occurring at the depths. In this way, material possessions gain social meaning not only because they have instrumental use in sustaining and developing our daily lives, but also because they function as symbols of identity, personality, and self-expression.[19]

Down at the very depths of our material experience, such enduring associations between people and things are not wholly designable. After all, for personal reasons, one can feel emotionally attached even to a turnip or a hubcap.[20] As Louise Crewe, a fashion theorist examining object meaning, placement, and value in domestic consumption, points out: "Questions of value, meaning, and worth remain mutable in time and space and open to interpretation."[21] Each user possesses a unique assemblage of memories and experiences, which render the user's perception and experience of each object as unique, vigorous symbols of the self, and carriers of great personal meaning and significance. At this point, it is important to note the obvious discrepancy between the things that matter and the things we use to communicate status, identity, and self to others. Too often, product life research conflates these two factors, leading to distorted results and false insights. Our most cherished possessions may very well play no part whatsoever in the external mediation of our identities, while the products we commonly deploy as signifiers of status do not necessarily "matter" to us in any significant way. The precious and few objects occupying the depths of material experience often play an entirely different role from those more abundant objects occupying the shallows. In contrast to the turbulent materialistic shallows, the deeper zone houses the treasure: the enduring possessions we cherish and keep safe for decades, even generations.

Just because an experience may be described as "shallow" does not mean it is negative, and conversely, just because an experience may be described as "deep" does not mean that it is necessarily positive. This kind of value judgment is a surprisingly common distraction for researchers examining the breadth of human experience. Instead, this depth metaphor simply provides a scalar means of thinking about experience based primarily on its degree of psychological resonance, rather than whether an experience may be described as positive or negative, good or bad.

Superstition and Belief

Superstition is a form of supernatural belief whereby we attribute an event to a force beyond scientific understanding or proof. Humans are naturally inclined toward supernatural beliefs. Many highly educated individuals experience a powerful sense that there are patterns, forces, energies, and entities operating in the world that are denied by science because they go beyond the boundaries of natural phenomena we currently understand.[22] Superstitious thinking is rife, particularly when it comes to our experiences of physical objects. Even self-confessed rational thinkers find their everyday actions influenced by irrational beliefs and their underlying conditions. Today, "superstition" is widely considered a derogatory term denoting beliefs or practices considered irrational. To be labeled "superstitious" (wrongly) implies you have little grip on reality, live your life in a way that is divorced from logical thought, misunderstand science or causality, and have a positive belief in fate or magic or fear of what is unknown. Owing to the term's pejorative implications, therefore, items considered to be superstitious are generally referred to as folk belief in folkloristics.[23]

Superstition has a certain survivalist tone in that it clearly derives from an attempt to attribute unlikely positive outcomes to repeatable behavioral patterns. Because of their entanglement with our survival instinct and sense making, these supernatural beliefs are both common and difficult to get rid of. Natural selection tends to generate far-reaching heuristics, or rules of thumb, to form patterns of cause and effect and, in so doing, make sense of the world and its causal relations. If making correct associations confers a strong survival advantage, then doing so will outweigh the negatives of making many incorrect, "superstitious" associations.[24] Despite the seemingly outlandish nature of superstitious thinking, we are all influenced to some extent by underlying beliefs in these kinds of supernatural forces. Many of us, for example, still engage in the common superstitious practice

Deeper Experiencing

of touching wood to avoid tempting fate, or choose not to walk under ladders. Superstition often refers to a religion not practiced by the majority of a given society, regardless of whether the prevailing religion contains alleged superstitions.[25] The term is also commonly applied to beliefs and practices surrounding luck, prophecy, and certain spiritual beings, particularly the belief that future events can be foretold by specific unrelated prior events.[26]

We can understand some of this in terms of the placebo effect. For example, engaging in superstitious practices surrounding healing can relieve anxiety and could have a significant positive physical effect. In the absence of advanced health care, such effects would play a relatively major role, thereby helping to explain the persistence and popularity of such practices.[27] Similarly, placing hope and faith in one's material possessions—like a copper coin that brings you luck, or a ceramic Buddha that wards off evil—also has a similar soothing effect. This might be referred to as a form of "magical thinking," in which people make false attributions about the causal relations between things in the world, a space in which our experiences of the physical and the metaphysical momentarily blend together. Of course, this may be likened to a form of apophenia in which people make meaningful connections between unrelated things, an "unmotivated seeing of connections accompanied by a specific feeling of abnormal meaningfulness" in relation to that connection.[28] This phenomenon, in many ways, links to a broader universal human tendency to seek patterns in random information through a kind of involuntary cause-and-effect reasoning process.

People who practice various forms of magical thinking do not require an explanatory theory behind their actions any more than the average person tries to grasp the pharmaceutical workings of aspirin,[29] as the prominent anthropologist Gilbert Lewis once declared. He highlights the double standard relating to certain phenomenological, lived experiences as opposed to others. Most of us accept the "scientific proof" behind aspirin without question, even though we do not understand how it works neurologically. In other words, what we believe about a given thing powerfully affects our experience of it, and the positive or negative nature of our experience influences the positive or negative character of our actions and their outcomes. In describing the relationships between exam performance and superstitious rituals, Stuart Vyse, a psychologist specializing in belief and superstition, explains how such rituals can reduce stress and thereby improve performance. He tells us how there is no magic, but there is a bit of "calming magic" in performing a ritualistic sequence before attempting a high-pressure activity. Any old ritual will do, says Vyse.[30] If you believe taking a lucky seashell to an exam will increase your performance, then it probably will—not because the seashell is lucky in and of itself, but because you are now more predisposed toward a positive mind-set because of your belief in the shell. This kind of belief is deeply subjective, and objects therefore possess as much power as you believe they possess. These objects have the power to influence performance by increasing people's self-confidence in their abilities; multiple studies have found a significant effect.[31] A charm is lucky if you wish it to be so. The influence of the charm depends on your belief in its inherent powers, and it is the belief itself that changes your fate, not the charm.

Transitional objects perform in this way. The term "transitional object" was coined in 1951 by Donald Winnicott as "a designation for any material to which an infant attributes a special meaning and by means of which the child is able to make the necessary shift from the earliest oral relationship with mother to genuine object-relationships."[32] Winnicott was an English psychoanalyst whose influential work helped shape understandings of object relations, particularly in young children. These kinds of objects help familiarize the otherwise unfamiliar situation and, in so doing, increase a person's capacity to effectively cope. Like anchors, transitional objects provide something steady and constant when all around appears to be shifting and unstable: a stake in the ground, a point of familiarity, an old friend materialized. In psychology, transitional objects are material things that help children make the emotional transition from dependence to independence: teddy bears, bottle tops, twigs; they can be anything at all. Sherry Turkle urges that most transitional objects have their "holding power" because of the particular moment and circumstance in which they come into one's life. She argues that many objects are naturally evocative because we associate them with times of transition, and these transitional times are rich with creative possibility.[33] Most readers are familiar with the American cartoonist Charles Schulz's character Linus and his blue security blanket. He drags it around wherever he goes, nibbling on its corner or curling up with it when the going gets tough. But children do have active imaginations that operate free from the constraints of scientific proof. For example, many children believe that icebergs and clouds are made for a purpose, that our lungs want to take in air when we breathe, and that the moon appears in the sky because it wants to see the child. However, this type of reasoning is thought to reflect a lack of understanding of biological and physical phenomena, and to be replaced by more mature thinking when the child becomes older and thus more familiar with the workings of the physical world.[34]

The ability to use transitional objects for self-soothing and comfort during stressful times is developed during infancy. This practice of deploying material things to satisfy immaterial needs by no means ends after childhood. Indeed, the use of transitional objects continues throughout a person's life, helping one to stabilize the ego during times of stress and uncertainty. In this way, transitional objects are not exclusive to children, and we continue to use them into adulthood, though not always in the form of security blankets. A recent survey by the hotel chain Travelodge, for example, found that just over a third of British adults sleep with teddy bears.[35] As adults, we use other types of objects that serve a similar purpose to transitional object used by children. It is common for adults to keep prized possessions owned by their parents when they were growing up, such as a watch purchased with your mother's first paycheck, an oil painting by a great-grandparent, or a scratched old brass fishing reel that landed your father's first salmon, long before you were even born. The specific example is not so important, but the principle is: if it mattered to them, then it matters to you. For adults, the notion of a transitional object may expand to include representations of one's family, home, and culture. These objects have significance for the person and give psychological strength by representing people's emotional attachments and the things, people, and places that matter to them. Objects are emotional. Whether a wedding ring, an heirloom dining table, or a tatty old teddy bear, objects can bring us a sense of comfort and connection to our past. Many adults consider the comfort and security that transitional objects provide as essential to their mental and emotional well-being and experience high levels of grief, panic, or remorse should these kinds of significant objects become damaged, lost, or stolen.[36] As you read this, you might even be reflecting on possessions that you cherish, and the different reasons why they matter so.

The label "superstitious" is not a neutral designation. Calling a belief or practice superstitious usually implies that the belief or practice is irrational, deluded, or at the least silly. People also have a sense that human progress has depended on humankind overcoming its superstitions and delusions, replacing magical explanations with rational ones, and consigning belief in the efficacy of good-luck charms to the dustbin of history.[37] However, superstitions persist, even in our technologically advanced world, with many of us still possessing lucky objects. Good-luck peanuts made their first appearance at the Space Flight Operations Facility at NASA's Jet Propulsion Laboratory (JPL) in 1964 during the Ranger 7 mission. JPL had six failures before this effort, so the pressure was on to succeed. The Ranger 7 launch day arrived, and with it came the peanuts. Ranger 7 performed flawlessly, as did its successors, Ranger 8 and 9.[38] In his yearly YouTube interview, President Obama spoke with the YouTube personality Ingrid Nilsen. She asked the president to bring a personal item and to speak about what it means to him. He brought several items: a rosary given to him by Pope Francis, a little Buddha figure given to him by a monk, a lucky poker chip given to him by a biker, a statuette of the Hindu god Hanuman, and a Coptic cross from Ethiopia.[39] All these objects are things that people have given to him over the years, during his travels around the world. During his presidency, these objects spent a great deal of time in his pockets, not perched on the shelves of his living room or displayed on the mantelpiece in the Oval Office. Each of these objects reminds him of specific values and people he has met along the way. They instill confidence in stressful situations and help him to believe in himself during times of adversity, as others have believed in themselves in similar times.

Objects laden with superstitious belief are wide-ranging and vary dramatically from person to person. However, the significance of such objects may be broken down into two distinct types: "culturally significant" and "personally significant." Often, culturally significant superstitions are associated with the most ordinary of household objects. Perhaps because they are in common circulation and always ready at hand, they tend to be written into a great many superstitious narratives and beliefs throughout the ages. Common household objects such as mirrors, salt, and scissors also provide a shared unit of exchange for people. They are things we all have, and so they provide ideal material from which superstitious cultural tropes may be cultivated, practiced, and shared. In Utah, the culturally superstitious practice of placing a rusty nail through a lemon is believed to ward off evil.[40] Doing so creates a talisman to protect against the evil eye and the misfortune it may bring. Talismans of this kind are common and appear in almost all cultures around the world. Although the form and their inherent stories may differ, the belief that these objects hold magical properties that bring good luck to the possessor or protect the possessor from evil or harm is commonly shared. In terms of the depths and shallows of material experience, these superstitious beliefs play a key role in determining the depth at which an object resides. Clearly, the hadal zone of our material world is filled with artifacts boasting more than a little superstitious belief or magical association. Indeed, the moment the nail pierces the lemon, two anonymous epipelagic objects become interlocked and are transformed into one singe hadal treasure. The plummet from the shallows to the depths can happen instantaneously, like the birth and formation of meaning itself. In this way, products do not always require a prolonged period to establish a depth of significance to us; material things can plummet to vast depths in a heartbeat.

Across the world, superstitious cultural practices abound, commonly with an everyday object cast in the leading role. Members of the Azande people of north central Africa believe that rubbing crocodile teeth on banana plants can invoke a fruitful crop. Because crocodile teeth are curved (like bananas) and grow back if they fall out, the Azande observe this similarity and want to impart this capacity of regeneration to their bananas.[41] To them, the rubbing constitutes a means of transference. In this case, the tooth itself is not the meaningful thing; any crocodile tooth will do. Rather, it is the ritual of rubbing the tooth on the leaf that is meaningful and symbolic. In Scotland, a pair of scissors must be handed to a friend with the blades facing back toward yourself, to avoid cutting (severing) the friendship. If you drop them, you must have somebody else pick them up for you; it is considered bad luck to pick them up yourself. In Denmark, people throw broken plates at their friends' and family's houses on New Year's Eve to bring good luck in the year to come. In Spain, citizens eat twelve green grapes and slip on a pair of red underpants at the stroke of midnight to achieve the same effect. In Russia, it is considered incredibly fortunate to be pooed on by a bird, and in Serbia, spilling water behind someone will bring them luck. These culturally superstitious practices are as much about the material objects as the ritualized way in which these objects are deployed. Collectively, they provide evidence of the pervasiveness of superstitious thinking in a skeptical age of evidence-based scientific certainty, which has all but displaced millennia-old bodies of supernatural belief and culturally meaningful materials practice.

In contrast, personally significant superstition and belief tend to manifest around highly idiosyncratic objects that are unique to each individual person. Like a language only one person can speak, these deeply uncommon object stories are nontransferable and fail to develop beyond personal significance into cultural significance. Such personally significant objects tend to be found in high-uncertainty situations such as gambling, war, sports, and test taking. In the case of sports, for example, research has found that superstitious behavior in top athletes positively correlates with the importance of the game, and negatively correlates with the degree of control an athlete feels at any given time.[42] Athletes in such situations need to feel as though they have some certainty or control. Furthermore, personally significant objects usually have a unique history, are sometimes gifts, are occasionally customized by the person, and are often lucky objects. During World War II, soldiers carried photographs of sweethearts and children with them on missions, as well as a variety of amulets, talismans, or special possessions like silver dollars, baby shoes, horseshoes, dolls, caricature figurines, and coins.[43] The basketball legend Michael Jordan wore his lucky University of North Carolina shorts underneath his Chicago Bulls shorts, and we have already learned of the personally significant objects Obama keeps in his pockets. And, of course, you and I also possess objects that we feel hold certain supernatural properties, be it luck, power, confidence, or connection to another person, place, or time that holds great personal significance to us.

Dark Objects

Most of the time, the objects we cherish hold a kind of energy that we consider to be positive. In contrast, we consistently distance ourselves from objects that possess negative and unwanted energies and associations—such is the power of our belief in the supernatural capacity of material things. Negative experiences are powerful. When negative associations form around an object, we feel compelled to distance ourselves from it, ejecting it from our lives. Nasty letters from former lovers or photographs of people who have hurt you typically fall into the category of a "dark object" we feel compelled to purge ourselves of. Of course, we will sometimes actively seek out negative experiences and even pay for them, for example, a disturbing Japanese horror film that leaves you feeling uncomfortable about the dark stairwell in your house for days. However, what distinguishes these types of experiences is that we choose them. Even as we interact with them, we remain somehow external to them, immune to any lasting effects.

We are all familiar with the belief that certain objects are negative and can bring misfortune or harbor bad luck. For example, fewer than 10 percent of condominiums across the United States have a thirteenth floor that is named as such. Builders avoid this number because of its associations with bad luck and instead replace it with numbers like 12b or 14a. The tweak is comparable to the way some Asian cultures view and deal with the number four. In Mandarin Chinese, the pronunciation of the word "four" (四) and "death" (死) is similar: sì for "four," sǐ for "death." This leads to a similar floor-naming struggle in parts of the word where Mandarin is commonly spoken, and buildings often skip levels containing a number four.[44] Beyond the shared cultural significance of certain unlucky numbers, dark objects are unique, one-off, and embody a more individualized form of negativity.

In the realm of material things, tales of cursed objects have flowed throughout the ages and continue today to weave threads of uncertainty through the material fabric of our lives. Take the story of "the cursed door," for example. Back in the sixties, an old door is salvaged from the debris of a recently demolished house. The door was cleaned up and reused in a family house. The night after the door was hung, weird things began to happen, like loud pounding coming from the cavity walls by the door. After several nights of fierce pounding, the homeowners were convinced the door was haunted and had it removed. Sure enough, the pounding stopped and never returned.[45] Whether these inanimate objects can be cursed, carry misfortune, or possess traces of their previous owner has yet to be proved either way. But the point of this discussion is that we *believe* they might do, to an extent, and this reasonable doubt is enough to make us wary of them. In the more rational parts of our minds, we feel sure that none of this is true; how can it be? However, on a more primitive level, we "half believe" these ideas, and this is enough to make us err on the side of caution and avoid these objects.

Sweat and blood may have DNA, but not bricks and mortar from a house. Rather, we sense something else in these objects, something supernatural. Bruce Hood, an experimental psychologist exploring the cognitive processes behind magical thinking in adults, describes the now-empty space on

25 Cromwell Street where a house once stood. The house belonged to the notorious serial killers Fred and Rosemary West and was the site of twelve known murders that took place between 1967 and 1987. The property was demolished, and since then nothing has been built in its place; the site is tainted. In October 1996, Gloucester City Council ordered the removal of all physical traces of the home where the Wests raped, tortured, and murdered young girls in the 1970s.[46] Even the paving slabs that lined the patio beneath which several bodies were buried were crushed into dust and scattered across several landfill sites in unmarked locations—such was the necessity of these dark objects being destroyed and removed from all memory.

Despite the extreme, horrific nature of the West house, the darkness that we experience in material things is more common than one might think. My old secondary school, for example, is a dark object to me. The discomfort I feel when driving past its gates is palpable, even now as an adult some thirty years later. Even writing these words feels a little uncomfortable, as it reactivates a broken part of me that still wanders the school's corridors, anxious and full of dread. On a more everyday level, dark objects reveal themselves frequently. At the market, my hand glides over the cheapest eggs to reach for the organic, free-range ones lying just beyond them on the shelf. I can almost feel the suffering pushing its way out of the clear plastic box. Just as a metal detector might alert its operator to the presence of ferrous material buried beneath the surface, my hand momentarily registers the systemic cruelty lurking beneath these dark objects, sold by the dozen.

The practice of removing "bad houses" and destroying their component parts is more common than one might imagine. It is also typical that no new houses will be built on the site. Instead, sites like this tend to be turned into utilitarian public spaces such as parking lots. This editing of our collective cultural memory is common and can easily be seen when we take a closer look at the way our public spaces are curated over time. We also see the inverse of this practice all around us. Just look at the way we memorialize and preserve material traces that are considered positive evidence of our past. We get rid of the objects that link us to an unwanted past, and we preserve and cherish the objects that link us to the past we prefer. The West house is a particularly extreme case, and few of us, thankfully, will ever have to deal with material encounters as harrowing as these. Nevertheless, the phenomena driving these examples are connected to milder forms of everyday object experience. All objects are "contaminated," in an experiential sense. They are not "clean," or quarantined from the world, and come infused with both wanted and unwanted associations and connotations. Weston Baxter leads research exploring the experiential dimensions of product ownership, in particular something he calls "contaminated interaction." He argues that object contamination arises in three ways: "hygienic contamination," "utility contamination," and "territorial contamination."[47] According to Baxter, "hygienic contamination" occurs when an object threatens a person's health or feeling of hygienic cleanliness. You might experience this when handed dirty cutlery in a restaurant or when sitting beside a train window with a greasy patch from someone's hair, which is now right next to your face. "Utility contamination" occurs when an object's functional attributes are damaged. This can happen when we update our operating system and find that certain older functions are no longer compatible. "Territorial contamination" results when objects seem to be marked by another person.

In our experience of objects, territorial contamination is surprisingly dominant in the way it makes or breaks our relationships with objects. As I have discussed, products can be positively or negatively contaminated, depending on their provenance. Even new products fresh out of the box come with contamination, depending on the values of the brand from which they emerged, or the current association one might have with that genre of object at a given moment in time. Despite their prominence, all these meaningful associations and beliefs are deeply unstable and liable to change, without warning, at a moment's notice. In this sense, the meaningful associations we form with designed things are deeply unstable and are powerfully influenced by both rational and irrational phenomena.

Rich Experience

As matter that we must negotiate, products shape our daily experience in ways that spark thoughts and experiences, and designers can therefore influence what these thoughts are. Yet although we may assign meaning to a given object or material, meaning cannot exist outside the body. It is within us, and our beliefs, that we find meaning. Objects, materials, or spaces cannot hold meaning in and of themselves; only our interpretation of these things will produce meaning. In this way, meaning draws from "lived experience"—what has happened to you up until this point—and is typically associated with sets of abstract relations and conditions, which create a lasting impression on us. The nature of this lived experience is both complex and poorly understood.

In design, experience is a process, not a thing; a verb, not a noun. Thinking of experiences as nouns is unhelpful. It artificially isolates them within their own abstract reality, divorced from context and any cumulative sequential meaning. That is to say, what one experienced moments before will influence the way one experiences in the present. I prefer the term "user experiencing," as it implies an ongoing action, a process. It also more effectively enables the splicing of parallel threads of experience, like using Facebook on a smartphone with a cracked screen while riding on a bus running late for a coffee with a friend you need to break difficult news to. We have very little control over the particularities of product experience, owing to their intertwined and cross-contaminated nature. I challenge any experience designer to untangle that one and present any kind of design direction that meaningfully contributes to such a complex series of entangled experiential processes. This example is not particularly dramatic and represents a typical experiential moment for many. In real time, experiences happen sequentially and can be understood as a cumulative stack of interrelated experiential touch points. Each of these touch points may, of course, be understood and examined as an experience in and of itself (the morning light flooding through the condensation-covered bus window, the rhythmic creaking of its rusted suspension). These isolated moments are the "nouns" of experience—the nameable things that occur when we encounter the world. On naming these isolated moments in lab-like isolation, we might be fooled into believing that we have a grasp of the discrete component parts of user experience. We can name these experiences, right, so we must therefore know and understand them. This is true to an extent, but the problem arises when you recognize that people do not encounter the world in insulated laboratory conditions.

Human experiencing is messy—a continually unfolding process, which cannot be chopped into nameable parts without a significant loss of meaning. Of course, boundaries exist to delineate one experience from the next, but these mechanistic boundaries are ontologically problematic, implying a toylike modularity to our being with the world, which simply does not exist in such reductionist terms. We must question the usefulness of such crude distinctions. Indeed, they may even be damaging; to establish the mythology that human experience can be constructed, brick by brick, like a child's Lego castle, is pure folly. One might argue that doing so creates the illusion of control and understanding, but the detachment from the temporally diffuse nature of our lived experience is stark. It also fails to recognize the ways in which each experience contaminates the other, rendering analysis of experience in quarantine moot. Rather, when it comes to understanding how we experience, think not of the singular experience in isolation but of the accumulation of experiences, dynamically working together to negotiate a collective, aggregate meaning.

Any business that claims its product is wholly positive is lying. Take education, for example. Even the best university degree in the world will involve a fair amount of negative emotion: the intellectual struggle, the demoralizing self-doubt, the anxiety of competing deadlines, the fear of failure. An experience always contains a blend of positive and negative, but in most cases, businesses tend only to promise the positive aspects of their products and services and blame any negative aspects on the customer. If a product fails to provide a positive emotional experience, most people raised in a consumer society will seek an alternative that does. Over the past several decades, the dominant commercial response to this behavior has been to design increasingly positive user experiences and strip out any trace of negativity from consumers' encounters with the material world. To ensure their own survival, businesses anxiously commission designers to maximize positive experience and eradicate negative ones wherever possible.

Experience designers specifically aim to optimize the subjective experiences users have with a product. Much of the work in this space rests on the fragile premise that positive experience is always desirable, and negative experience is always undesirable. However, if we look at examples of rich and meaningful human experiences, we quickly realize that these assumptions are not always valid.[48] Design's primary approach to the design of experiences can be alarmingly prosaic, given the potential depths of human experiencing that can be probed and explored through richer and more complex encounters with material things. The range of possibilities considered by most experience designers is alarmingly narrow. Designers typically consider an unnecessarily limited set of emotions for the product experiences they orchestrate. Experience designers should feel more confident in developing user experiences that more closely mimic the kinds of mixed emotions and blended experiences we live out in our daily lives.

Indeed, if you want to develop product experiences that are deeper and richer, you need to go beyond pleasant, one-dimensional encounters. As Hassenzahl suggests, product experiences should be "worthwhile" or "valuable" to avoid the pitfall of shallow amusement in experience design.[49] Designers commonly assume that we only want to experience positive emotions and experiences, and that we will go to great length to avoid their negative counterparts. This is largely untrue, although we do have an inbuilt tendency toward the positive. In contrast, designers can enrich product experience by explicitly involving negative emotions in the user-product interaction. Negative emotions have been shown to contribute to pleasurable experiences in other domains (e.g., computer games, movies, art),[50] but negative emotions are generally designed out of human-product interactions.

Typical examples of people actively seeking so-called negative emotions include riding roller coasters, parachute jumping, or watching horror movies. In all these examples, the negative emotions are wanted, so they are distinct in that regard. You want to experience the exhilaration of the ride or the terror of the movie, which you know will be short-lived, and kind of nice when it is over. But aside from these edge cases, we are instinctively hardwired to be compelled toward things that make us feel positive emotions (e.g., happiness, satisfaction, pride), and repelled by things that make us feel negative emotions (e.g., disgust, shame, anger). We see this human characteristic play out in all kinds of ways. Our photo albums tend to strip out the negative memories and instead memorialize only the memories we want to revisit and hold on to. Similarly, circles of friends tend also to be edited in a way that maximizes positive emotions and experiences and minimizes the negative. Through routine yet unconscious maintenance, we construct a personalized version of our own reality with a significant bias toward the positive. And so it is human to want to inhabit a more positive emotional world. Yet the world is not like this. Continually convincing people that it is misaligns our expectations with reality to the extent that we become ill-equipped to handle even the slightest dissonance or error in our lives. Through a century of design's blind pursuit of positivity, we now find ourselves trained out of knowing how to cope with even the slightest degree of cognitive dissonance. Today we see negative emotion as something wrong and thus to be avoided. Of course, some designed experiences elicit negative emotion in ways that are simply insulting, racist, or gender biased, for example. These forms of unintended negativity expose underlining biases within the people who created them. The presence of designed objects eliciting such experiences is deeply unpleasant for the people subject to their oppressive powers; such objects have no place in this world. And so I am not calling for a universalizing "open season" on negative emotion, regardless of its moral or ethical position. Far from it. Rather, what I am advocating here is a widening of the frame when it comes to designing experience-rich product interactions.

From the perspective of making, particularly hand making, error is a collaborator, an ever-present colleague whose virtual hand pushes yours, when making and doing real things, in the real world, in real time. Error tends not to happen in the mind; things usually work perfectly well up there. It is only when an idea or intention is shared with others or applied in practice that its incongruencies may emerge, when the idea "rubs" against reality in a way that exposes previously concealed properties and characteristics. The interaction designer Mary Tsai argues for an illuminating reframing of error, set against modernity's trajectory of certainty, toward which design and technology blindly hurtle. She describes how AI has us mesmerized with its own perfection, compelled by the feverish pursuit of a future free from failure and imperfection—a highly certain and controlled future that, as we know, does not exist.[51] Error is natural and manifests through a perceivable discrepancy between expectation and reality. As a subjectively determined phenomenon, error is most commonly understood as something negative, evidence of imperfection, or a problem to be steered away from. When working with mechanical or computational techniques, designers tend to be particularly error averse and rarely embrace materials that contain flaws or imperfections, partially because of designers' expectations of the machine's ability to produce flawless outcomes.

Aversion to error is understandable. If we think of error in the context of a high-stakes situation like landing an aircraft or performing an AI-assisted hysterectomy, it becomes abundantly clear that predictable, error-free control is vital. Yet, in many cases, error contributes to the richness of product experience. Oftentimes it is the idiosyncratic nature of error that gives character to products, lifting them from anonymous mass-produced homogeneity to something greater. We see this in things such as air bubbles in handblown glass or knots in an oak table. Many designers work at the threshold of error, to celebrate such partially controlled processes, working at the limits of a machine's ability and forcing it to fail, thus producing extraordinary results. We see this in products like the Tel Aviv–based product designer Shira Keret's *Monolith*, which comprises a collection of water-jet-cut marble vessels. Water-jet cutting is a computer numerical control (CNC) machining process that uses high-pressure streams of water to penetrate material, in this case marble. As the water jet cuts deeper into the material, the jet loses both pressure and precision, which causes an uneven line of cut. This unevenness introduces variability, such as subtle fluctuations in the density of the material or minor irregularities in water pressure. The resulting cut is "wavy," not straight, and this wave tells a story. Although the product was formed through mathematical precision and mechanical creation, it can highlight the faults of the technology used, as well as the limitations of using an organic component such as water as a cutting tool. Work like this poses a timely challenge to the idea that computationally driven design and manufacture equal higher levels of precision and control. They really do not have to, and when agency, or "free will," is given to the machine, extraordinary characteristics are revealed that tell us about the material, the machine, time, gravity, physics, and the world. Without error, we engineer these discrepancies out of the product and, in so doing, erase 90 percent of the product's story—usually the most interesting 90 percent.

Recent interest in the value of negative emotions can be understood as a countermovement against a postwar Western trend to overvalue optimism, hopeful thinking, and positive emotions, especially in the United States. In her book *Bright-Sided*, Barbara Ehrenreich argues that this positive-thinking ideology has been attempting to ban the expression and experience of negative emotion, as though it were toxic or harmful, from American daily life.[52] Anyone involved in the pursuit of designing "meaningful stuff" should be wary of this blind prioritizing of the positive above all else; it is an uncritical domain of false hope and broken promises. Life is messy, and the meaningful associations we form with material things are hopelessly entangled with all of it—good and bad. Designers have developed an unhealthy preoccupation with positive emotions and wholly constructive product experience. All negativity is painstakingly edited out of our interactions with designed products and services, leaving only the wanted experiences. The level of artifice here is jaw-dropping, a contrived and polished version of reality in which negative emotions have no place. Should they arise, these unwanted emotional responses are experienced as a problem to be solved, a flaw to be fixed. This form of design is reminiscent of an excessively hands-on parenting style, wherein the parents cannot bear to see their children experience boredom, even for a moment, even though boredom can be an important thing to experience, a challenge to think your way out of. On the contrary, design should build on the richness of our everyday experiences, not demolish it bit by bit to leave behind the smallest kernel of short-lived happiness.

Steven Fokkinga is a design researcher who advocates a more expanded user experience paradigm that embraces both positive and negative emotion. He argues for a kind of dynamic, emotionally plural mode of interaction design. His work explores the design conditions under which both positive and negative emotions may be embraced and even be considered favorable. Imagine you are moving to a new city, says Fokkinga. What emotions would you experience? You might feel sad about leaving your family and friends. At the same time, you might also feel hopeful about the opportunities awaiting your new life, joy over the prospect of exploring your new city and meeting new people, and anxiety about not knowing anyone yet. The combination of all these different emotions makes the transition a complex but "rich experience," as Fokkinga calls it, that you will long remember.[53] Our interactions with the designed world are cognized in this way, splicing together threads of positive and negative emotion to form a complete, rich experience. Examples of mixed emotions, where we experience both positive and negative emotions in parallel, are common. Though we might describe such positive and negative emotions as occupying opposite ends of a theoretical scale, they are complementary. As the sixteenth-century theologian and philosopher Pierre Charron remarked: "Pleasure, and pain, though directly opposite, are contrived to be constant companions."[54]

Although rich experiences are commonplace in the art and entertainment industries, they are virtually absent in the domain of consumer products.[55] The broad spectrum of human emotion provides design with a diverse repertoire of material to work with when formulating deeper realms of product experiencing. The truths and the fictions, the facts and the beliefs, the positive and the negative, all coalesce to form the totality of a product experience. Moreover, negative emotions do not necessarily lead to unpleasant experiences. In fact, negative emotions often contribute to the formation of deeper product experiences. Even the experience of repairing something like an old dining table, taking it from a broken state (negative) to a functioning state (positive), can still offer a rich experience, when that experience is considered longitudinally, as a whole process. Despite an obvious migration from negative experience to positive experience, the experience as a whole process, unfolding over time, requires healthy measures of both positive and negative emotion to develop in the way it does. Indeed, seeing the world through material engagements involves recognizing what Tim Ingold describes as the dynamic flows of interaction by which meaningful relationships emerge rather than develop as planned.[56] In this scenario, the material is like an actor; it can assume many different personalities, depending on the role it is asked to play.[57]

Interaction design can create a "protective frame" to transform a negative user emotion into a rich user experience. A protective frame is a theory from clinical psychology, used to explain how people need to be "shielded" from the unpleasant elements of negative emotion using a protective frame.[58] According to Fokkinga, any negative emotion, when combined with a protective frame, can form the basis of a rich experience. Negative emotions can make a positive contribution to user-product interactions because of their unique effects on human perception, attitude, and experience. Michael Apter, a behavioral psychologist who established reversal theory, points to the dynamic qualities of experience and how people regularly flip between opposing psychological states. Apter introduces the concept of the protective frame: negative emotions that are generally experienced as unpleasant can become pleasant when they are experienced within a protective frame, an evaluation of a situation in which the individual frames the threat or trauma as manageable and therefore transformable into excitement.[59] Apter describes three distinct types of protective frame: the "confidence frame" (parachutists can enjoy the jump given their confidence that the equipment is fully operational); the "safety zone frame" (one can enjoy a cliff as long as one is sufficiently distant from the edge or behind a solid fence); and the "detachment frame" (viewers of a scary movie can enjoy the thrill because they know what they are watching is fiction).[60] All protective frames are based on creating experiential distance between oneself and the emotional stimulus. According to Apter, the "distances" these different protective frames provide are in some cases physical, like an actual distance or barrier between the individual and the threat, but can also be psychological, like the notion that a threat is not real or can be dealt with.[61]

When we think about product experiencing, we see the designer's role expand: designing not only for commercial, market-led contexts but also for broader societal ones. Our work becomes less about designing product experiences that can be passively consumed and used today, and more about exploring contentious, imaginary experiential situations that might exist in years to come. And it asks us to imagine not only things we desire and want but also undesirable and unwanted things.[62] To ignore this unwanted hemisphere of human emotion is to deny our very humanity; to persist in blindly promoting positive emotional experience at all costs is to push toward an artificial future with no place left for sentient people. By expanding our frame, design can widen its experiential scope to encompass a greater emotional range and, in so doing, more firmly grasp the wide-ranging experiential conditions of human flourishing.

5 Aging Spectacularly

Improving with Age

We have been raised to believe that the quality of products deteriorates over time, and material things generally get worse the longer you have them. Indeed, the functional lives of objects have shortened consistently since the 1950s, but this is not to say that things get worse over time, or there is no place in the world for things that last. The sense that something special or unexpected must have happened for a thing to last is peculiar—that the object must somehow be exceptional or have drifted laterally into an alternative mode of keeping, in which only a select handful of artifacts reside. To think that something unexpected must have happened for a manufactured object to have made the transition from "product" to "possession"— from something external and other to something enduring and entangled within one's complex unfolding self—is a bizarre situation for design to find itself in.

But what does it mean to "endure"? We hear of products that "survive against the odds" or "stand the test of time." What "odds" are these, and what "test" are we referring to exactly? It suggests that the default position of reality is that of destruction, decay, and failure. Wholly fabricated, these "odds" and "tests" are a way for us to collectively imagine that the world is set against the survival of material things, products are designed to fail, and disposal is more a question of when than if. Because we are beings whose only certainty in life is death, perhaps this projection of our own mortality on the objects we surround ourselves with is to be expected or even natural. While this pattern of thinking does carry a certain existential tone, the question persists: What is being endured here, anyway? What are these objects surviving against or in spite of? Woven throughout these questions are deep assumptions surrounding the corrosive, decaying forces of the environment and, perhaps more crucially, our unfailing ability to outgrow the material conditions that surround us. For many, the dominant critique of a material world overflowing with short-life products is that things are no longer designed to last. Things do last, just not as long as they used to. Note that when I say "last" here, I am referring to longevity in use, not an object's material longevity as discarded matter decomposing in the depths of some landfill. That is certainly not the kind of "lasting" I am arguing for.

Different types of endurance manifest across diverse settings and contexts. One thing that needs disentangling here is the use of the word "endurance" to describe organisms, as opposed to objects. Commonly, the two terms are used interchangeably, and while they bear strong parallels, the differences are important to isolate and consider. In living, sentient beings, endurance most commonly refers to "will." In this scenario, endurance describes a situation in which an organism can exert itself for a long period of time. With this kind of emotional endurance, we might be challenged by the psychological pressure a given situation confronts us with. This form of endurance requires us to motivate ourselves and "work at it" to prevail, for example, to endure a dull meeting or to have the endurance to babysit three mischievous toddlers. With objects, endurance tends not to be about "will" but instead refers to an ability to withstand harsh and challenging external forces. These forces might be environmental (e.g., moisture, temperature, light) or cultural (e.g., aesthetics, trend, taste). These modes of endurance align well with typologies of functional and psychological obsolescence and show the extent to which object endurance and obsolescence are tied. Of course, the human body does have physical endurance characteristics and is a hard-wearing, long-lasting organic structure capable of growth, regeneration, and self-repair. In the case of objects, however, we may not say that "will" forms a part of their enduring traits. Despite rapid advances in artificial intelligence and machine learning, the sensation of an object possessing its own free will is highly subjective and in no way represents what is going on at an objective level; we perceive autonomy, but it is not really there.

Endurance is a state detached from time in a quantifiable sense. That is to say, the endurance of a given thing—be it a sprawling city or a single moth pupa—is subjectively measured against how long that thing ordinarily lasts. For example, a lacewing can endure a sudden tropical downpour lasting ninety seconds, while the Colosseum in Rome may be said to endure millennia of environmental wear and tear—not to mention the four million tourists trampling all over it each year. In common parlance, "endurance" captures all this sufferance, resilience, fortitude, and hardiness. But in the end, the assessment of whether something possesses properties of endurance or not will always be based on the observed tendencies of how long the thing tends to last.

But what role does the process of "wear and tear" play in all of this? Products are materialized moments in time. Whether you like it or not, they age. If you look around, everything that your eyes fall on ages, from the glass in the windows to the concrete of the building you can see through them. All of this is changing, all the time. Of course, our experience of the everyday tends to happen through a series of fleeting glimpses, which provide a fragmented, artificial portrayal of reality. These passing snapshots capture isolated moments in a far longer and more complex timeline of an object, material, or building, for example. Only through sustained engagement with a given thing—be it a physical thing (e.g., pen, armchair, house) or a digital thing (e.g., app, web browser, operating system)—can we begin to understand it in the lengthier context of change, over time. Hannah Arendt, hailed as one of the most influential political philosophers of the twentieth century, describes the symphony played between user and used, consumer and product. She speaks of how "use contains an element of consumption, in so far as the wearing out process comes about through the contact of the use object with the living, consuming organism, and the closer the contact between the body and the used thing, the more plausible will an equation of the two appear."[1] In this way, wear is revealing in the way it speaks to the specific character of relations sustained between the person and the product over time.

As if to prove our supremacy over natural laws, we fabricate the made world as though it can be fixed, set in place, and frozen in time. In doing so, we form impossible expectations of permanence, of things that last for centuries. To transcend the inevitability of age, we have fabricated an alien world of durable metals, polymers, and composite materials. Immune to the glare of biological decay, these materials grossly outlive our desire for them. Even the level of value assigned to ideas and theories relates directly to their longevity and how well they have stood the test of time. John

Wood, emeritus professor of design at Goldsmiths College, University of London, describes how this coincides with a popular idea of rigidity as a paradigm, and how we use material metaphors such as "concrete," "ironclad," and "material" to elevate the status of thoughts and opinions.[2]

When designing for positive aging, it is tempting to specify robust, scratch-resist polymers or atmospherically resilient alloys as a tactic to extend product longevity. Yet we find that it is often the perceived fragility of a thing that leads us to care for it, pay closer attention, and keep it in active service longer. In this way, a product designed for a century of use should seek not only to be beautiful but also to be beautifully fragile. This encourages a form of care and consideration in relation to how we use, store, and maintain a given thing. It needs us, just as we need it. One might argue that it is exactly because of this fragility that common objects demand our attention, sensitivity, and care—so that perceived fragility may be a critical component in the design of meaningful stuff. Let us not forget that even a cheap, disposable plastic drinking cup has a material life of several centuries already. It is already physically durable, and landfills the world over are stuffed full of durable products just like this. In contrast, the plastic cup's useful life covers a handful of minutes at best: a short life indeed for such a materially robust product. Of course, this is less of an issue when using materials that have not been engineered out of the biosphere, into what might best be called the technosphere. Arendt describes how a wooden chair rides the tide of biomass to arrive back at its point of origin. She tells how "if left to itself, or discarded from the human world, the chair will again become wood, and the wood will decay and return to the soil from which the tree sprang before it was cut off to become the material on which to work and with which to build."[3]

This mismatch between physical and experiential durability plays out on the grandest of scales and is not just confined to small household objects. Whether the drained, algae-coated diving pool from the 2016 Summer Olympics in Rio de Janeiro or the vine-entangled, graffiti-scrawled luge and bobsled track from the 1984 Winter Olympics in Sarajevo, objects befall disuse—sometimes all too quickly. Although these are extreme examples of single-use products, they demonstrate the same disposable, throwaway design mind-set that underpins the creation of everyday disposable products like paper cups and diapers. These highly durable things are incapable of maintaining relevance outside of their initial contexts and the changeable, unstable conditions behind their creation. These stadiums now stand as unintentional monuments of concrete, glass, and steel, attesting to the fickle nature of our collective attentions. Clearly a dissonance exists between how long things last and how long we meaningfully engage with them.

Matter in Motion

Although subjective responses to product aging vary, people commonly recognize when, and to what degree, aging has taken place. It is commonly accepted that aging surfaces can add character to objects, giving them a history and a story. Yet, often, this process devalues products by enforcing their sense of age and consequent loss of currency. We intuitively understand what brand-new, lightly used, and heavily worn look like, because of the material traces left behind. After a product is released from the atmospherically controlled, dust-free factory conditions, marks on the product's surface offer up evidence of an object having been in the world—that it has encountered its environment, occupied its habitat, and has lived. Prominent aesthetic changes caused by minor wear and damage to otherwise pristine enclosures (patina), combined with incremental upgrades in hardware and regular tweaks to styling, contribute to the rapid turnover of material goods.[4] Minor changes to the look and feel of new products shorten the life expectancies of products we already own by pushing perceived obsolescence and wasteful behaviors.

Product testing of electronic devices by manufacturers typically focuses on avoidance of functional failure,[5] not gradual wear and tear, and certainly not the way the object's exterior skin might change over time. This approach is peculiar because, from the user's point of view, the latter point is of great importance. Researchers at Loughborough University's Closed Loop Emotionally Valuable E-waste Recovery project argue that the kinds of degradation phenomena discussed here may be split into two distinct processes: "wear" and "damage." Wear refers to the kinds of aging that inevitably occur through protracted periods of product use, handling, and carrying. Through even careful use, materials will gradually polish, mark, and scuff. Over time, these marks signal that a product has been well used, but not necessarily abused or mistreated. Think of the way the matte finish on the corners of a phone case becomes polished through repeatedly being taken in and out of the pocket. Damage, on the other hand, refers to material changes occurring from less careful use and storage. While these material changes may not necessarily affect product performance, they do signal a carelessness in the way the product has been owned. Consider carrying a phone in your pocket with keys or frequently dropping it on hard surfaces.[6] While this sort of damage rarely impedes functionality, it writes new tales of the user's clumsiness and ham-fisted approach to things onto the surface of the product. What is more, these newly scripted tales of imperfection are out there for all to see: your clumsiness on display, so to speak.

Aging frequently sounds the death knell for consumer products. Indeed, "Imperfection is not a usual aim within the context of industrialized product design. Under general norms, products are manufactured as clones of a 'perfect' original and product surfaces are prized for their 'perfect' flawless state."[7] However, we also see examples to the contrary, products that are elevated to a higher level, largely through their design-led capability to age gracefully. Designed by Kravitz Design for Leica, the Correspondent is one such example—albeit an expensive one, with the limited-edition model retailing at $24,500. Leica Camera AG is a maker of premium-quality cameras, with almost legendary status to photography enthusiasts through the sheer quality and precision of the company's products. Leica's high-end special edition camera is an extension of the 2003 Leica MP, which stands for "mechanical perfection." One would assume that notions of mechanical perfection, operational precision, and excellence in material quality would not sit well alongside worn glossy black enamel paint, eroded and polished brass corners, and furrowed, crinkled snake leather. Extending this notion of timeless quality, original design features like the scalloped focusing ring and knurled aperture dial connect this camera to its 1950s predecessor. This shows how Leica's deep technical expertise and product build quality have withstood the test of time, despite the deluge of digital cameras. The subtle promise being made here is that Leica has been right for this long and will likely continue to be right for decades to come. Although only a small number of Correspondents were produced, they held center stage in Leica boutique stores. Leica describes how the product has been built for long life and lasting value. The company celebrates how "after years of hard use, when a bit of bright brass begins to show through, . . . the camera and its owner have clearly shared many memorable experiences."[8] The hand-aged design is an attempt to bring back the aesthetics of an old, beautifully aged

Leica, the kind that might have been handed down to you by a parent after decades of use, a story in progress for which you are now coauthor.

And so products can be designed to age gracefully, even spectacularly. From the worn-down front doorstep to the key-chipped paint surrounding its lock, aging material surfaces reveal signs of life by scripting the user within the otherwise inanimate object. Patina is seldom associated with an increase in value, and users generally perceive marks, scratches, and scrapes as negative occurrences. Yet when products are designed to improve with age, they challenge such fixed ideas. In the case of objects that we have owned and interacted with over time, our association with these marks will often be positive because of our intimate understanding of how they were acquired. This is particularly true of damage, which may be interpreted very differently depending on when it happens: a scratch caused by dropping a new product differs in meaning from a similar scratch caused by an interesting event in the product owner's life.[9] Whether deliberate or unintentional, every crack or scratch that materials manifest as we interact with objects inscribes a story.[10] These range from complex tales of intense use and shared adventure to disappointing accounts of an object's deteriorating quality and failure to endure. Of course, these forms of patination have little to do with actual material resilience or durability. Rather, they reflect a societal preoccupation with what an appropriate condition is for certain typologies of materials and objects to be in. The stories that patina writes are outwardly projected on those around us, not just mirrored back to the user; they are social. It is not just *that* an object has aged but rather *how* its aging is experienced and understood by users and their peers. Patina therefore provides clues of use and other decipherable indications of an object's otherwise secret life. The scrapes, blemishes, and imperfections afforded through patination have the power to make or break our relationships with products.

Imperfection has a key part to play in cultivating meaningful product experience, because of the way it grabs our attention. Plumbing is a bit like this. When it works perfectly, we do not notice the pressurized pipes quietly coursing water beneath our floorboards and within the dark cavities of our walls. When plumbing fails, however, we immediately become aware of its presence. Like an awakening, product failure jolts us into awareness and exposes us to its previously concealed characteristics and potentialities. Although this form of imperfection may not be so desirable, it does provide a compelling example of how perfectly obedient products and systems can easily be forgotten or overlooked by the user. Giving value to imperfection, and particularly giving value to imperfect material qualities, leads to a reconsideration of the relationship we have with everyday products.[11] The striving for seamless interaction, at this point, seems a somewhat misguided pursuit that sorely overlooks much of the rich experiential potential that material things have to offer. Perfection, in this way, not only is boring but also sets up unrealistic expectations that practically guarantee users will be disappointed down the line.

In traditional Japanese aesthetics, wabi-sabi is a worldview centered on the acceptance of transience and imperfection.[12] Wabi is the quality of a rustic yet refined solitary beauty. Sabi is the trait, be it the green corrosion of bronze or the pattern of moss and lichen on wood and stone, that comes with weathering and age.[13] With this, a certain love of roughness is involved, behind which lurks a hidden beauty. This beauty, with inner implications, is not a beauty displayed before the viewer by its creator. Rather, a piece will lead viewers to draw beauty out of it for themselves.[14] This aesthetic is sometimes described as one of beauty that is imperfect, impermanent, and incomplete. We see it in certain styles of Japanese pottery, which are often rustic and simple looking, with shapes that are not quite symmetrical, and colors or textures that appear to emphasize an unrefined or simple style.[15] To discover wabi-sabi is to experience a form of beauty in something that at first glance appears decrepit, worn, and ugly. It is an understated beauty that exists in the modest, rustic, imperfect, or even decayed, an aesthetic sensibility that finds a melancholic beauty in the impermanence of all things.[16] This element of "time passing" is inherent in the design: a chipped vase, a faded piece of cloth, a rogue air bubble trapped in a pane of glass. These moments of material imperfection are gifts. They elevate the status of the object by connecting us to a wider and more expanded set of universal principles, the physics and dynamic environmental realities of our world.

According to Japanese legend (ca. 1522), a young man named Sen no Rikyū sought to learn the elaborate set of customs known as the Way of Tea. He went to tea master Takeeno Jōō, who tested the younger man by asking him to tend the garden. Rikyū cleaned up debris and raked the ground until it was perfect, then scrutinized the immaculate garden. Before presenting his work to the master, he shook a cherry tree, causing a few flowers to spill randomly onto the ground. To this day, the Japanese revere Rikyū as one who understood to his very core a deep cultural thread known as wabi-sabi.[17]

As a design sensibility, wabi-sabi helps bring our attention closer to a more authentic materials experience by acknowledging three simple realities: nothing lasts, nothing is finished, and nothing is perfect.[18] This aesthetic offers a sharp contrast to more conventional design ideologies of perfection through symmetry and uniform proportion, which themselves are shaped by an appetite for control and mastery over one's environment. Importantly, these diametrically opposed aesthetic values should not be superficially labeled as "crass West versus sophisticated East." Too often we hear this romanticism propagated by scholars in the so-called West, yet it is a false opposition that has little or no basis in reality. Edward Said's cautionary term "Orientalism" warns against these Western constructions of the East as exotic, mysterious, and enticing. Said was a Palestinian American scholar who founded the academic field of postcolonial studies. He argued that these constructions are not merely passive descriptions of other people and places but are both central and harmful to the imaginative production of those places.[19] To take this idea further, we might even question the validity of terms like "East" and "West" in today's globalized world. What does this even mean, and to what extent is it helpful to clumsily split the world and its people in half? When we lazily toss these terms around, what is often meant is more a "west and the rest" form of rhetoric, in which ideas of the West persist in their assumed sociocultural dominance over all other ways of life.

Let us shift scale for a moment. Consider the land we walk on, and our shifting experience of it as it changes, as we too change. As a dynamic assemblage of matter—earth, rock, water, wood, and so on—the landscape may be understood as an enduring record of the lives and works of past generations who have dwelt within it and left there something of themselves.[20] As matter in motion, we may understand landscape as an emergent property of its conditions, continually "becoming" in response to human and nonhuman forms of engagement, over time. In the example of landscape, we encounter a more spatially expanded form of patination, spanning kilometers, not millimeters; millennia, not months. Stretched across a longer time horizon, landscapes shift in response to human and nonhuman acts of use. Some of these shifts show quickly (e.g., a house being built on a hilltop), whereas others are relatively slow, sometimes taking centuries to appear (e.g., the changing course of a winding river). Landscape, therefore, is a product, semihandmade, held in a perpetual state of becoming, or as Nietzsche might put it, "a chaotic world in perpetual change and becoming."[21]

Temporally speaking, users operate in a present that is always "infused," and which they are further infusing, with pasts and futures. In Husserl's subject-oriented language, these pasts and futures are referred to as "retentions" and "protensions."[22] Our experience of the present moment constitutes an infusion of such retentions and protensions. The French philosopher and public intellectual Maurice Merleau-Ponty describes our experience of one moment to the next as something we do not conceive or notice as an onlooker. He tells us how the present is not marked off from a past that it has replaced or a future that will, in turn, replace it; rather, it gathers the past and future into itself, like refractions in a crystal ball, as it continues to "become" over time.[23] To Merleau-Ponty, embodiment, perception, and ontology are entangled in this way and as such are inseparable.

We are caught in a constant state of becoming. In this continually unfolding state, we do encounter products, materials, and spaces as fixed and unchanging entities. Nietzsche defines such isolated and static entities as false concepts that are the necessary mistakes that consciousness and language employ to interpret the chaos of this state of becoming. Rather, when we engage with objects, we encounter a far more dynamic assemblage of matter in motion. Objects also exist in a social sphere where they transform social relations and are transformed by them in terms of perception and sometimes the physical form. What is more, not only do objects change in material ways—a scratch here or a dent there—but more important, the ways in which we individually read and understand these objects change also, as we continue to unfold, to become. Indeed, though these materials may endure, products occupy relatively fleeting moments in time. As ephemeral as bubbles, these brief object experiences are enchanting and short-lived. No sooner have we grasped their qualities and meanings than they adapt into something else. In resilience thinking, this innate capacity to absorb disturbance and accept change, rather than defensively resist and block it, is key to survival. In evolutionary biology, it is not the strongest species that survive, or the most intelligent, but the most responsive to change.[24] In the designed world, this is sorely misunderstood, and designers generally consider the ever-present tension that exists between states of change and stability to indicate that the two states are at odds with each other, when, in fact, they are not.

Stuff Is Spatiotemporally Diffuse

The fumbled idea that products exist within clearly contextualized moments in time, encountered by rational users, is pure folly. Critical designer and educator Anthony Dunne, for example, calls for visions of the future "that reflect the complex, troubled people we are, rather than the easily satisfied consumers and users we are supposed to be."[25] Material objects are spatiotemporal in their context and deceptively distributed across space and time, a complex hybrid of material and energy. They bring with them their own stories, which collide awkwardly with our own histories and futures. This collision comes to categorize the moment of encounter, the user experience, by which the subject works to reconcile a storm of meaning circulating around the object itself.

Products, and their materials, are spatiotemporally diffuse. As a result, so too are our experiences of them. Their contexts are continually shifting, held in eternal flux; our experiences of the material world are never quite the same and exist in constant flow. The notion that "everything flows" derives from Heraclitean thought, which likens the flow of a river to the eternal flow of change over time.[26] In this analogy, the materials, objects, and environments we inhabit represent a continually shifting spatiotemporal assemblage, in perpetual flux. Heraclitus was a pre-Socratic Greek philosopher commonly known for his doctrine of change being central to all things in the universe: "You cannot step twice into the same river," and "All things move and nothing remains still." The notion that everything flows is highly relevant for design—a field preoccupied with unwieldy universal principles, fixed outmoded doctrines, and enduring obsolete creative processes and intellectual frameworks. It implies the passing of time while also using the metaphor of water to show such change and motion at work.[27] Heraclitus reminds us that in our pursuit of permanence, we place ourselves fundamentally at odds with the most essential underlying principle of the world: change. Indeed, Heraclitus's river itself could be described as a different river from moment to moment, since the water flowing in it is different from moment to moment. Or to put it another way, a river may be understood as a process, not a thing.

Designing for uncertainty, and a world caught in perpetual flux, requires a fresh creative focus on flexible, dynamic, and adaptive products and systems, a focus on designing products with the capacity to change as the world around them changes. These protean systems reframe products as dynamic and unfolding processes, rather than static and fixed things. In creating products capable of adapting to meet changing circumstances, we avert generating enormous quantities of waste. This notion of a material thing also being a dynamic process is not as bizarre as it might at first sound. For example, children constantly change—their interests, ideas, dislikes, confidence, and so on. Yet we are predisposed to accept and expect this aspect of the process of aging and development. Importantly, like the changing Heraclitean river, the child does not become a different child with each change. Rather, the child has changed and adapted in some way, making the child a slightly evolved version of what he or she was before. The child remains the same in some respects, and not the same in others. Product change, in this way, is a cumulative, ongoing process that builds incrementally over time.

Timothy Morton is an object-oriented philosopher engaging with issues relating to human effects on the environment. In metaphysics, "object-oriented ontology" (OOO) is a twenty-first-century Heidegger-influenced notion that—unlike Kantian anthropocentrism—rejects the privileging of human existence over the existence of other, nonhuman objects.[28] In this way, Morton refers to what he calls "hyperobjects." He argues that hyperobjects are vastly distributed in time and space, relative to humans, and are both seen and unseen. He tells us that "a hyperobject could be a black hole. A hyperobject could be the Lago Agrio oil field in Ecuador, or the Florida Everglades. A hyperobject could be the biosphere, or the Solar System. A hyperobject could be the very long-lasting product of human manufacture, such as Styrofoam or plastic bags, or the sum of all the whirring machinery of capitalism."[29] Hyperobjects are sprawling spatiotemporal ideas, so vast that they defeat traditional ideas of what "a thing" is. These forms of "object systems" are all around us, yet as designers and users, we overlook their presence. Morton's notion of hyperobjects invites us to consider a more expanded role for design in steering system-level change and to reconceive our creative responsibilities as spanning much longer horizons of time. Though spatiotemporally distributed, hyperobjects still have "form"—albeit within the social imaginaries we assemble to conceptualize their complex systemic conditions. This form may be likened to the form of an idea, a family, or an experience. It lacks tangibility but can also be understood as a whole, unified entity as opposed to an uncoordinated collection of random fragments. The Finnish architect Eliel Saarinen argued that we should always design a thing by considering it in its next larger context—a chair in a room, a room in a house, a house in an environment, an environment in a city plan.[30] Doing so helps designers more easily reflect on the multiple systemic conditions within which designed things simultaneously reside.

Despite their hard, molded profiles and well-defined material boundaries, objects do not exist as static, isolated entities. They are diffuse assemblages of matter, energy, and value, spatiotemporally connected to a host of living and nonliving systems. They are emergent properties of the systemic conditions from which they originate. In the same way that smart objects do more than just enable a world of increased corporate control and surveillance, they also provide the tools to expose and reorder the very logics and procedures that created them.[31] Morton argues that through this process of exposure, the fallibility of our dominant sociotechnical system comes sharply into view. When one learns to see the system behind the object, a more complete picture of the distributed world we inhabit begins to appear.

Taken at the level of human experience, Morton's contention is hardly surprising. Several "spatiotemporal threads" run throughout our experience of the everyday, forging meaningful associations within the dynamic and unstable environments of time, place, objects, and materials.[32] In this way, the lived experience of "things" is spatiotemporally situated—a splicing together of multiple spatiotemporal threads, which, when encountered as an ontological whole, shape our experience of the designed world. This whole experience is constituted by continuous engagement with the world through acts of sense making at many levels. It is "continuous" in that we can never be outside of experience, and "active" in that it is an engagement of a concerned feeling, self-acting with and through materials and tools.[33] These spatiotemporal threads reminds us that experiences are particular in time and space; they are context dependent and relate to a particular person, in a particular situation, at a particular time. These myriad particularities are unstable and continually in flux. Therefore no two experiences will ever be identical. For example, seeing the same movie in the same cinema for a second time is always going to elicit a different experience.[34]

As a purposive activity, design is both intentional and strategic.[35] Yet, often, what we consider to be "designed" is only one small part of a broader object story—the tip of the iceberg, if you will. The creative disciplines rarely acknowledge this. The American ecologist Garrett Hardin refers to the diffuse, unpredictable nature of all human action in his First Law of Human Ecology. He says: "We can never do merely one thing. Any intrusion into nature has numerous effects, many of which are unpredictable."[36] When we design an everyday electronic device such as an LED flashlight, designers will consciously attend to issues relating to aesthetics, ergonomics, material quality, performance, cost, and so on. Designers consciously attend to these considerations, and as a result, the outcome—in this case, a flashlight—is shaped by the thinking that underpinned its formation. However, a wider shaping also occurs because of this "design (non)thinking" of a less intentional nature, such as the poisoning of groundwater through toxic compounds and heavy metals found in the gloss black paint. In this way, objects and their associated social and environmental impacts are all designed, albeit unwittingly. They manifest as an aggregate of consequences, accrued through decades of purposeful decisions, on the one hand, and apathetic nonengagement with the world, on the other. The design choices we make, and do not make, contribute to their diffuse formation in ways that are problematic in their consequences. Indeed, no designer ever set out to make the world a worse place, though designers may inadvertently have done so. The designed world is the way it is because of the thinking, values, and understanding that underpinned its formation; the things we choose to focus on and think about; and the myriad other things we do not.

Classics as Temporally Anchored Nodes

I would be remiss to write a book on design that lasts without even briefly discussing "design classics." After all, they represent a form of design that has, indeed, lasted. Classics reside within the sociocultural imagination. They connect us to people, places, and times that continue to hold value and persist in their ability to garner shared, meaningful associations across large populations. The notion of "the classic" flows through practically all forms of art and culture and has a varied attribution, from a work of art, literature, or film to designed objects like cars, bottles, and furniture. The term "classic" is used to valorize things that can be judged over an extended period to be of high quality, to exemplify innovation in their type, and to possess lasting cultural and symbolic value. It takes time for an object to become recognized as a classic. Whatever lasting impact the design has had on society, together with its influence on later designs, plays a large role in determining whether something becomes a design classic or not. Thus design classics are often strikingly simple, going to the essence, and are often described with words like "iconic," "revolutionary," "valuable," or "meaningful."[37] What is often referred to as "classic" would perhaps be more accurately described as "iconic"—a historical object that we recognize and value because it symbolizes a particular time, person, or brand.

It is well understood that design classics are works that have withstood the test of time and remained immune to the passing whims of an ever-unfolding present. Many design classics are still in production today as exact reproductions of the original design, manufactured and retailed under license. Others iterate on the typology that the original classic design established, such as the wooden clothes peg, the disposable ballpoint pen, or the bicycle. Crossing cultures and time spans, classics set new standards in their field and inspire other designers and manufacturers. In this way, classics also have a degree of universality, which supposedly widens their appeal and scope for participation. While classics themselves do not change, the ways in which they are reinterpreted on the basis of a changing world frequently do.

In identifying universal characteristics of classics, several factors are consistently present: "quality," "impact," "recognition," and "permanence." Classics possess high quality, both in their material and in their design and construction, and normally continue to represent the best in their class, despite not being the latest model. Designed for Alfonso Bialetti, the grandfather of the industrial design luminary Alberto Alessi, the Bialetti Moka coffee maker (ca. 1950) exemplifies this enduring quality, with its iconic eight-sided form and high-quality aluminum body. Classics are also game-changing objects that made a significant impact. This might include a breakthrough in manufacturing process capability or a positive disruption of long-established norms and behaviors. For example, Mini was a car designed out of necessity, coming about because of restrictions in fuel supply during the 1950s caused by the Suez crisis. At this time, the automotive designer Alec Issigonis was tasked with designing a car that was more frugal than the large cars of the day. The Mini was revolutionary at the time and quickly became a British icon that influenced generations of car designers to follow. Classics must receive "recognition" of their status to hold the title. The American poet, novelist, and editor George Parsons Lathrop writes in an introduction to Hawthorne's *The Scarlet Letter*: "We speak of a book as a classic when it has gained a place for itself in our culture, and has consequently become a part of our educational experience."[38] Finally, classics demonstrate permanence and have stood the test of time. One might argue that much of the magic of engaging with classic objects comes from the way they transport us to past times and places. By engaging with them, we find ourselves momentarily transplanted in times long since passed. Part of how classics achieve such permanence is through being reinterpreted and reapplied as the years unfold. The Swiss Army knife from Victorinox (1891) is a folding pocket knife that can also open tins, saw branches, and repair mechanical objects. Later iterations on the knife added more components (e.g., nail files, sewing eyes, fish scalers, tweezers, magnifying glasses). The device is so widespread today that it has even entered common parlance as a metaphor for adaptability and usefulness. Every age sees its own experience in a classic and views it through its own individually polished lens. One may argue that for a classic to be designated as such, it must possess each one of these characteristics (quality, impact, recognition, permanence) in abundance.

Yet despite the fanfare surrounding classics, they have a concealed, darker side that warrants discussion here. Classics can hold us back, anchoring thought in long-extinct social, ecological, political, and economic conditions. Unless carefully understood, design classics are the "Make America Great Again" of design: an overly nostalgic affection for times that no longer exist. We must, of course, learn from these past times, and even respect them, but our design intent must never be to try to re-create the past. Design education itself has a somewhat unhealthy preoccupation with the "design classic." Lecture theaters crammed with undergraduate students endure weeks and months of nostalgic rhetoric on classics, supposedly training future designers to think like past designers or somehow learn from their design moves. Is this really the education we want? As Cameron Tonkinwise, professor of design studies at the University of Technology Sydney, decries: "Many designers get their aesthetic education from the contemplation of such 'museumed objects,' silhouetted out from their background everyday life, and recast in the ethereal neutrality of the photographic studio."[39] Classic after classic crosses the professor's projector screen, a stream of fetishized objects from a world that no longer exists, narrated with a dewy-eyed adoration for how things were but no longer are. Pasts and futures fold into the present, to inform and shape the design mind-set. In this way, to engage with what has come before is essential; no design education is complete without this element. Design has influentially shaped objects, spaces, and systems over time. Classics offer one means of punctuating this long, continuing story, though their importance as exemplars is regularly overemphasized. Discussions of design classics, when properly historicized, can help us get a sense of design's evolutionary arc, but at design school, this is seldom how the discussion is couched. Classics have a way of becoming part of the shared experience of a whole culture, through which a collective body of aesthetic values and taste regimes unwittingly takes form. To persist, this recognition must be maintained through frequently repeated transmission. A classic, in this way, is understood as classic because we are educated to believe that it is—an established and prebaked sociocultural appreciated that we inherit, rather than one we create for ourselves. A degree of consensus must be reached, of course, whereby an established base of sociocultural agreement encircles the product and its status as a classic. This consensus must be propagated and maintained to endure. The classic, therefore, is like a story that survives through the retelling.

We often hear the label "timeless classic" affectionately slapped onto culturally significant objects, things that have somehow persisted in remaining relevant despite changing times. However, this notion of timelessness is grossly misleading. What people intend when using this term is well meant: that an object has somehow held on to its perceived cultural value despite the passing of time. Yet our experience of classic objects is not timeless but "time-full." These objects connect us to the times and conditions from which they emerged in ways that are not always sufficiently acknowledged; they transport us back. Classics are temporal nodes anchored in time. Pursuit of ahistorical and timeless "design classics" is a perilous endeavor for those looking to design products that last. The very attempt of making objects immortal, of making them simply last "longer" rather than last "meaningfully" in their sheer perfection and everlasting completeness, casts them outside of time, making them somewhat unrelatable as things.[40]

Instead, we must design material things to last longer through their innate ability to change over time: things that are not finished and can be repaired and altered. This goal requires a new design philosophy of things that are deliberately incomplete, and things that stay in motion. Whether designers, and the brands they work for, can create things that are not finished remains to be seen. This requires a letting go, a relinquishing of control and an invitation to the user to coproduce the designed world, and not just interact politely with it.

Whether in a vast and sprawling landscape or in a pocket-sized mechanical device, promoting greater degrees of user agency and coauthorship blurs the threshold between acts of design and acts of use. This can be troubling for designers, as it destabilizes the often tightly held notion of design's absolute authorial claim, set against the relative impotence of the passive consumer. The Italian philosopher Giorgio Agamben describes this overlapping territory as a "zone of indifference" wherein the process of designing extends into the process of using.[41] At this critical juncture, use itself becomes a vital form of making, and the user is promoted to a coproducer of the object. Importantly, in considering use as making, we reframe users as active performers of open-ended, unfolding, and improvisational tasks.[42] Use, in this sense, becomes a creative and productive act. We can then see an important shift happening, whereby the object's openness, responsiveness, and hackability are functional to a democratic displacement of design authorship. Amplifying the agency of users through the object's open-endedness widens the user's sphere of action. It permits spontaneity, improvisation, and an enhanced mode of object interaction and engagement, rather than imposing on users the need to abide by a predetermined code of practice. This turns the object into a sort of "blank canvas" of use, a space for emergent properties to manifest in response to idiosyncratic patterns of use.

Why is it so peculiar for something to last today? We often hear our elders lament that things are no longer built to last. Their sentiments often originate from a place of deep-seated frustration about the failure of modern products to endure. They are not talking about a breakdown of emotional attachment or some kind of "looming cognitive dissonance" that agitates them out of ownership and toward wastefulness. Rather, these words come from a place of objective concern about the increasingly short-term, disposable character of modern products. What we also hear through these words is a higher-level critique of the ever-shortening life spans of material goods, exacerbated by a mounting sense of alienation from such "newfangled," unfamiliar objects. Of course, this is exacerbated by a lack of access to affordable and accessible outlets for repair, and a now decades-old trend in industrial design toward products that cannot be repaired in the first place.

It is easy to understand why we might covet products that have stood the test of time. Sure, things are not made the way they used to be, but then again, they never have been. Though coming from a well-meaning place of concern about the perceived drop in quality or artisanship, these sentiments are partially rose-tinted; changes bring both positive and negative consequences. Only when we compare the new version with an earlier one can we critically reflect on these positives and negativities and tease out points of difference that we like or dislike between old and new ways of doing things. Ordinarily, new systems are evaluated against older ones. Certainly, for individuals who have lived through a prior system and now grapple with the new one, their experience of the new is colored by their memories and subjective experience of the former, more familiar systemic condition. For people for whom the current system is the one they grew up in—the only one they have known—things are less complicated. The current system is easier to accept openly, and its strengths are more easily identifiable. In time, though, this system will mutate into something else, and at that point you might find yourself comparing it with the old one, the one you know now. This is how iteration and adaptation work. For this reason, I suggest that it is unhelpful to heed the nostalgic words of people longing for a prior system. Yes, we can reflect and learn from our shared design histories, and this reflection may help us better understand our present, but we cannot, and must not, go back.

The discourse on classics is riddled with contradiction and rests on fragile theoretical premises. Much of the debate assumes that a sole designer creates classics—a kind of "magnum opus" artifact. Objects in this narrow category do exist (e.g., an Eames lounge chair, a Philippe Starck lemon squeezer), but they are the exception rather than the rule. More frequently, classics exist as typologies of objects iterated on over time by multiple designers across multiple times (e.g., angle poise lamps, QWERTY keyboards). Furthermore, the assumed "universality" of design classics is a misnomer. That is, design classics tend to derive from a limited handful of design cultures, the United States, the United Kingdom, Germany, Italy, and Spain being common birthplaces of design classics. In a more global context, we might question the value of such colonial models of design history, drawing its points of reference from such a culturally limited palette of examples.

So what can we learn from classics to support design that lasts? Can future classics even be designed, or does an object achieve classic status purely by chance? Surely the push for future classics runs in opposition to the evidence toward longer-lasting products. Objects must not be so tightly tethered to social and cultural conditions, which in themselves are continually shifting and morphing into new things. It is nonsensical to umbilically link a fixed and static material thing to an unfixed and dynamic sociocultural thing. As one half of this partnership shifts, the other is tipped into obsolescence. I cannot think of the number of times that well-intentioned designers have approached me and requested the recipe for the "secret sauce." The design formula they seek would supposedly help them create the ultimate "future classic." They forensically dismantle classic products in search of the secret to their success, in the hope that their essence can somehow be extracted and then incorporated to new products. This sorely misses the point. It takes time to become a classic. If there is one definition of this abused term, it is the cognate ability of a given thing to resist the inevitable ebb and flow of taste and fashion. Classics cannot be invented; they evolve and have to win approval and slowly acquire value over time.[43]

Many classic designs are striking to look at but terrible to use. For the clear majority, these objects live only in the pages of books, magazines, and lecture theaters, so you would never know. You may also encounter classics through documentaries extolling the virtues of design geniuses through the ages. However, most people will never actually get to use these things, and so the secret of their poor performance remains concealed. What is more, many so-called classic products become waste for this exact reason: they are tethered to sociocultural conditions that we no longer wish to be associated with. They are promptly jettisoned from our material worlds as we strive to distance ourselves from their negative associations. Classics offer such points of punctuation and help us to navigate the temporal landscape of the designed world. Classics encapsulate "times" in powerful ways, and this encapsulation is what we experience when we engage with them.

Classics are an emergent property of temporal, cultural, economic, and political conditions. A consensus is reached as to their classic status, offering an enduring sociocultural nod to their sustained relevance and value

Aging Spectacularly

through the ages. The classic provides a point of entry into this assemblage of conditions, allowing us to engage with the particularities of that past, as distinct from our overly familiar present. A certain nostalgia encircles the world of classics, which is simultaneously understandable and troubling. They offer brief moments of reassurance in a world of acceleration, change, and ever-increasing levels of complexity. The social anxiety induced through this persistent acceleration leads one to wonder whether there is a form of reassurance in classics. This may seem like a cynical remark, but I find something deeply conservative about the backward-glancing emphasis placed on classics, motivated by a dystopian outlook on the future and an underlying dissatisfaction with the present. In a fluid, adaptive world, classics signal that not everything changes, and some values remain, and thus a degree of security can be found in their fixed nature. For some, these object encounters may be likened to bumping into an old friend. Such encounters offer a quiet reassurance that some things do not change, and there are still some things that last.

Aging and the Digital

Whether deliberate or not, every crack, scratch, and dent that products bear as we interact with them inscribe a story. These interactions with material things result in alterations, imperfections, and ultimately unique objects, which carry traces of time and life.[44] But what does this process mean for the digital artifact? The digital is caught in a constant and iterative state of renewal. Like a shedding snake, its old skin effortlessly peels away to reveal a fresh, glossy layer beneath. Digital objects, like apps or websites, have a capacity to continually evolve and change in a way that physical objects cannot. They are wholly customizable, upgradeable, repairable, and adaptable. They do not objectively age, unless you code them to, and these updates can be distributed around the world in the blink of an eye, with little or no material throughput. Before we get any further into this, a bit on the word "digital," which is used here as a shorthand, referring to forms of complex electronic technologies that generate, store, and process data. "Digital" is widely used to describe a plethora of interactions within a world becoming increasingly digitally mediated. In industrial design, we often hear of "the digital" relating more to the hardware that renders the data tangible, rather than the data themselves. This includes digital electronics comprising electronic circuits that operate using digital signals, digital cameras that capture and store digital images, or a digital computer that handles information represented by discrete values. Essentially, a series of "digital [insert noun here]" phrases could easily follow, but hopefully I have made the point that the underlying idea of the digital operates more as a conditional prefix than any clearly defined thing in and of itself. This is, of course, understandable. Our experience of the digital is largely shaped by the materials and devices that deliver it to the realm of human experience.

Even in the supposedly frictionless digital domain, a heavy reliance on materials persists. Without the material, the digital is still there, but from the perspective of human experience, it is kind of not there, as it remains beyond the scope of our conscious awareness. Without material things, you cannot really do anything with the digital or engage with it in any way, like the fabled Invisible Man, standing silently in the corner of the bathroom, his presence revealed only when the steam from the shower condenses on his raincoat and spectacles to expose his silhouette. Nick Foster, head of design at Google X, notes the false opposition behind terms like "physical" and "digital," and the role of material objects in a world increasingly dominated by artificial intelligence. According to Fosta, imagining this requires "a wholesale restructuring of everything we currently associate with objects: their affordances, what they mean, how they work, and who owns them." He argues that we have long been told that "software is the key to the future, but software always needs to be delivered through a 'thing.' Even if the future of software is ephemeral or audiocentric, it will still require electronic i/o (input/output) such as a speaker and a microphone, clustered together and housed in a 'thing.'"[45] Similarly, the specter of the digital requires material to render itself visible. Indeed, no experiential connection with the digital can exist without the presence of material things. Edge cases exist that initially appear to suggest the contrary. Augmented reality (AR), for example, simulates the presence of material goods in a home environment that remains otherwise minimal. Through the screen, we experience our hallways and living rooms filled with coveted design pieces—a Vitra armchair here, a Kartell lamp there. Within the AR-mediated illusion, the objects are there, and not there, all at once. Lower the screen, and you return with a jolt to the underwhelming reality of your own life and shoddy possessions.

It is now redundant for design to continue perpetuating the binary distinction between the digital and the analogue. Instead we should think of products as hybrid, comprising both analog and digital dimensions and capacities. Web developers versus product designers might naturally resort to such divisive terminology and consider their professional practice as occupying either wholly digital (intangible) or wholly material (tangible) worlds. After all, the skill sets required to operate on the coding end of the spectrum, versus the form-giving end, are so diametrically opposed that one can forgive the artificial separation. Users, on the other hand, do not draw such a distinction. To them, it is a much more holistic, entangled, and messy set of encounters. The environments in which experience unfolds are not static or fixed. They are deceptively distributed across space and time and exist as complex assemblages of analogue and digital, material and energy. We might call these "hybrid environments." Figuratively speaking, several braided threads flow through our experience of these environments. They are braided in that we experience them initially as an interwoven whole—a complete and total experience. Yet environments are poorly understood when examined at this holistic metalevel. To design for environments, we must first unpick the "braid," so that we might examine its numerous and distinct threads. Only then can we design experientially rich interactions with hybrid objects and environments. This term better serves the context and helps shift focus toward the crossbred nature of the digitally mediated object,[46] and the myriad blended worlds these objects simultaneously inhabit.

But can the digital threads of a hybrid experience really age? To address this question, we must distinguish between material and immaterial components. While the digital data (code) do not age, the electronic technologies (products) that generate, store, and process those data certainly do. Let us stick with digital data for a little longer. Digital data are data that represent other forms of data using specific machine language systems (e.g., C++, Python, JavaScript, Visual Basic). These languages can be interpreted by various technologies in various ways. Hundreds of programming languages exist today, and most programmers become fluent in only a handful of them. Despite their diversity, one thing all these languages share is their resistance to decay. In other words, code is impervious to the changing world around it; it does not corrode or become worn through use. Code can, of course, become messy, cluttered, and out of place, when updated by coders who lack inputting discipline or care. However, this would be more an issue of maintenance than aging per se.

The aging of digital experiences can only be discussed in relation to the other, newer experiences that surround them. Even drops in performance of digital experience can largely be attributed to the acceleration of the things around them, rather than any drop on the products side. Some tech products do slow down and are even designed to do so as more advanced, "heavier" operating systems are periodically released. However, most things only appear to have slowed, owing to the relative acceleration of the newer things that now run alongside them. And so, paradoxically, although the object has not aged at all, it does appear old. In this way, digital aging only occurs at the subjective level, and in relation to the changing things around it. This is a form of relative aging. The script remains unchanged to the last keystroke, the swarms of web-optimized GIFs and JPEGs lie dormant in their folders, awaiting activation, just as they did before. Nothing has changed. However, from the perspective of the user, everything has changed. The web page looks old, dated, and aged somehow. This subjective experience of aging is driven by the fact that the context around the thing has changed while the thing itself has not.

The "Diderot effect" describes this unstable, relativist character of our relationships with objects. We are drawn to things that reflect our unfolding identities. It therefore follows that the products one acquires will be somehow complementary to one another, as each object has been selected based on a similar set of identity-matching criteria. However, as the anthropologist and scholar of consumption patterns Grant McCracken points out, the introduction of new possessions can disrupt the previously stable collection of complementary products.

The Diderot effect helps explain the process whereby a new purchase or gift creates dissatisfaction with existing possessions, triggering a spiraling pattern of consumption with hugely destructive environmental, psychological, and social implications. The phenomenon is named after the French philosopher, art critic, and writer Denis Diderot, who first described the effect in his 1769 essay "Regrets on Parting with My Old Dressing Gown." In this short text, Diderot wittily describes a scarlet dressing gown that he received as a gift. At first, he was thrilled to receive such a beautiful garment. Soon, though, the presence of this new item made all his other possessions seem relatively shabby and out of date. They could not live up to the elegance and style of his new possession. He began updating all his possessions in a vain attempt to bring them up to the standard of the dressing gown, plunging him into huge debt, with no requisite increase in happiness or satisfaction. He writes: "I was absolute master of my old dressing gown, but I have become a slave to my new one."[47] This upward creep of desire,[48] as the prominent sociologist Juliet Schor puts it, is insatiable and has grave environmental and social consequences. No sooner have old things been updated than they themselves need replacing and updating, leading to an endless sequence of desire and disappointment.

6 Urban Mines

Owned but Not Used

The pursuit of design that lasts steers consumption patterns away from wasteful short-termism toward more resource-efficient, lasting modes of engagement with the designed world. Yet even if we can make this shift and design a world in which products endure, an underlying problem remains: the dominant consumption paradigm is still one of ownership, isolation, and possessiveness. The renowned political scientist C. B. Macpherson argued vehemently against the rising culture of possessive individualism he observed in affluent Western societies during the mid-twentieth century. In *The Political Theory of Possessive Individualism*, he confronts this "destructive social condition in which individuals conceive themselves as the sole proprietor of their skills and possessions and owe nothing to society for them."[1] Macpherson's powerful treatise points to a form of narcissistic isolationism in which life's primary purpose is to acquire and consume, to hoard and covet. In this flawed mental model, possessions become the prime unit of value and signifier of status. It follows, therefore, that one's success and one's assets are corequisite. From an outsider's perspective, the more you appear to "have," the better you appear to "be." Through this process of acquisition, each of us shores up an island of worldly goods, through which we might experience reassuring sensations of being skilled or having competency. This stockpiling of goods on our "private islands" is a common antisocial practice. From the user's point of view, it can only really be rationalized through comparing behaviors: how a person might be doing in relation to his or her neighbor, close friend, or older sister, for example. In this way, it is not always about "having" but more frequently a means of avoiding the antagonism of "not having," or at least of avoiding having less than those around you appear to have. The material throughput required to support the consumptive process underpinning possessive individualism is several factors beyond the carrying capacity of Earth; such acute levels of conspicuous consumption simply cannot continue. Today we find ourselves in a world of marginally more sustainable products. But if people continue to be socially conditioned into thinking they must own one of everything, these minor design improvements at the product level do not make a whole lot of difference.

I once came across an article titled "50 Things Every Man Should Own to Win at Life."[2] Troubling title aside, items in the list included a good umbrella, a tool set and workbench, a classy pen, a quality watch, a signature cologne, a scotch (whiskey) collection, and so forth. Of course, some of the "fifty things" are things I already have, or would like to have, but this is beside the point. What is troubling is the way the article is framed: the way it draws a direct correlation between having material possessions and social success (winning at life); the way it unconsciously points to the consumptive underpinnings of such status-seeking behavior and how to superficially project it to those around you; the way it propagates a prescriptive model of masculinity that chastises us into mindless consumption in a vain attempt to conform with someone else's tired caricature of what a man should look, act, and smell like. Of course, this is a "culture" we have created, and the author of the piece is just riffing off that. As prescriptive as a hipster's barbershop in Brooklyn, such prescriptive typologies of style and taste are common. These "taste regimes" shape patterns of consumption and waste.[3] This is one powerful way in which taste provides a mechanism for perpetuating social mythologies and hierarchical structures. Through sequential acts of consumption, stretched out over time, we sign up for one or more of these regimes. As the American novelist Chuck Palahniuk warns, "We find ourselves chasing cars and clothes, working jobs we hate so we can buy shit we do not need."[4] Indeed, that nobody is paying attention to your overpriced watch, or even cares what brand of cufflinks you wear, is lost on most people.

Due in large part to the dominance of possessive individualism, the material intensity of modern life has escalated to perverse proportions. Take a typical suburban residential street in the United States. Let us say there are sixty small family homes on this street; half on one side, half on the other. Now, instead of thinking of each building as a house, think of it as a container, in which you will find at least one of everything. Filled over time, these containers are piled high with stuff—overfilled drawers that no longer close, closets piled so precariously high with possessions we dare not even open them. One might speculate that this street, with its sixty houses, also has sixty power drills, sixty washing machines, sixty vacuum cleaners, sixty shoe-polishing kits, sixty lawn mowers, and so on. Zoom out from this street, and you will see it nested among a neighborhood of one hundred near-identical streets, each one also comprising about sixty houses. That is six thousand houses, each one a container stuffed full of near-identical products. Zoom out farther still, and the neighborhood becomes a city of twenty equally sized neighborhoods, with upward of 120,000 homes. That is 120,000 containers, which together store 120,000 power drills, 120,000 washing machines, 120,000 vacuum cleaners, and so on. "Ah, but not every household owns a drill," you might observe. Yes, that is true. But many households own several drills. My father, for example, happens to own three drills: one in active use, and two that he replaced years ago but would rather not part with, since they still technically work, though not especially well. Now, consider all the material things in a typical home—in every drawer and every closet. The average US household contains around 300,000 individual products, from paper clips and table lamps to ironing boards and screwdrivers.[5]

Relative to the rest of the world, we do live in a culture of excess in the global North. Children in the United States, for example, make up 3.7 percent of children on the planet but have 47 percent of all toys and children's books.[6] In addition to our materially swamped living spaces, the basements and attics of our homes are also crammed full of material things. Many of these are "just-in-case possessions," lying dormant, ready to spring into action should the moment arise. But, of course, it seldom does. In the United Kingdom, 30 percent of our clothes are unworn. This amounts to about £5 billion of unused clothing in wardrobes in any given year.[7] Rachel Botsman is a Trust Fellow at Oxford University's Saïd Business School. Her work explores the behavioral intersections of collaboration, trust, and technology. Describing the underuse of privately owned goods, she tells of how the average lawn mower is used for four hours a year, and the average power drill is used for only twenty minutes in its entire life span.[8] In my house, there are three of us, and we each have a bicycle. I used mine for about three hours last year; my son's bike was ridden a good deal more than mine, but still not much more than about thirty minutes per week on average. Products like bicycles spend most of their time lying in wait, just in case.

One might naturally assume that a house filled with such dormant, underused possessions might be the direct result of hoarding, overconsumption, or greed—the result of someone blindly buying too much stuff. Occasionally this is the case. However, many unwanted objects spend a period of "conscience time" in drawers, attics, and garages while we decide whether to dispose of them or not.[9] Storage spaces can often provide the time and distance we need to make the right decision about what to do with possessions we no longer want. The human geographer Jen Owen explores the hidden life of belongings during their time in storage, which acts like a kind of limbo for material things. She specifically refers to this storage phase in a product's life as "possession purgatory."[10] Her illuminating work shows how sites of storage (e.g., drawers, wardrobes, attics, self-storage units) provide locations for delaying the divestment fate (ridding) of possessions that we feel unready to part with. For example, her work exploring the relationships between possessions and bereavement shows how renting a self-storage space can provide a temporary solution for mourners at a time of loss and grief—a time when it is simply too painful to think through what to do with the personal effects of the deceased. These remote spaces are deliberately out of sight and out of mind. They allow possessions to be reencountered in a new, neutral context at a later date and under less desperate, emotionally turbulent circumstances. As a result, these storage spaces support practices of mourning by "distancing and delaying engagement with the possessions of the deceased, and the memories they trigger."[11] In this way, storage spaces—particularly those more removed from daily life—provide the necessary conditions for the final stage of an object's life cycle to take place in a considered and reflective way.[12]

Mining the Anthropocene

When products are relegated to the boxes, drawers, and loft spaces of our lives, their contexts change dramatically. Pulled out of service, these outcast objects sit idly in storage, like inmates on death row, awaiting their inevitable fate. Oftentimes possessions may sit in storage for a decade or more before their disposal, during which time the precious materials tied up in them sit idly out of reach. The Royal Society of Chemistry has been investigating how much old technology a typical household in the United Kingdom has stashed away. The initial survey revealed that half of UK households had at least one unused electronic device, and 45 percent of homes had between two and five.[13] The researchers speculate that as many as forty million digital products lie dormant in drawers and cupboards up and down the country, though this is a conservative estimate.

Storing underused material things can be counterproductive, especially when viewed through the lens of resource efficiency. In the case of an underused app, an album no longer played, or an archive of old emails, the material consequences of disuse are negligible. However, as described earlier in the book, when one considers the resources locked inside our material possessions, the consequences of idle products become far more serious. These material-rich "urban mines" have been stocked—albeit unintentionally—through decades of material consumption, importing rich repertoires of material into cities from some of the most far-flung corners of the world. Located in every affluent city of the world, these urban mines contain far higher concentrations of precious materials than the actual subterranean mines themselves; from advanced polymers and natural fibers to rare earth metals like tin, tungsten, tantalum, and gold, our modern cities hold rich reserves of material wealth.

As an industrial means of extracting geological materials from the earth, mining ordinarily comes with considerable negative social and environmental impacts. It takes place in geographically dispersed locations, relying on complex logistical systems to bring those myriad materials together into a single product. As many of our precious materials, like tin, copper, and zinc, become scarce, mining resources from beneath the ground becomes less profitable. In the emerging age of the circular economy, subterranean mining is increasingly considered a costly, outmoded industrial process. One of the most powerful things about reframing stockpiled and underused possessions as an "urban mine" (rather than "waste") is that all materials within these mines are local. In other words, we already have all the materials we need, right beneath our feet. Instead of materials residing in unprocessed states among many thousands of tons of rock and mud, they now sit perfectly processed, riddled in among the chassis and circuitry of our unwanted stuff. The dystopian imaginary of the urban plan I described earlier in the chapter—with its houses, streets, and neighborhoods stuffed with underused products—now takes on the character of a plentiful urban mine: a resource-rich seam of material wealth, sitting above ground, right where we need it. Established, outmoded industrial processes have us ravaging the earth and its people in search of the precious resources needed to manufacture our materials and products. Meanwhile we sit ignorantly atop stockpiles of untold material wealth, stashed away in the mine-like recesses of our homes.

The notion of the urban mine takes us closer to the vision of cosmopolitan localism: the theory and practice of interregional and planetwide networking between place-based communities that share knowledge, technology and resources.[14] It suggests a new social, political, cultural, economic, and technological "settlement" that could help address many of the twenty-first century's wicked problems.[15] As Gideon Kossoff eloquently states: "In a cosmopolitan-localist system, we can have attachments, commitments, loyalties, and a sense of belonging at multiple levels of scale; to our locales, other locales, and the planet as a whole."[16] Kossoff is a social ecologist and design professor at Carnegie Mellon University. His research focuses on the relationships between humans, the natural environment, and the designed world as the foundation for a sustainable society. His vision of "cosmopolitan localism" is powerful in the way it supports new ways of designing that catalyze the societal shift toward sustainable lifestyles that are simultaneously place based (local) and internationally networked (global).[17] The challenge, says Kossoff, is to help restore and reinvent households, neighborhoods, cities, and regions by enabling their inhabitants to recover control over the satisfaction of their needs and by redesigning satisfiers so that they are synergistic and place based.

Today, our designation of materials as "local" or "not local" is wholly inadequate, if a little conservative. Many environmentalists forcefully advocate the use of locally sourced materials: native species of timber, or wool shorn from locally reared sheep, for example. Yet much of the thinking underpinning these distinctions is centuries old and certainly falls short of usefully assessing the regional availability of a given material in modern times. Often, when people refer to a material as being "local," what they really mean is indigenous, native, or "of this place." However, when it comes to identifying and specifying local materials as a resource efficiency measure, this outmoded framing is profoundly unhelpful. For designers to discriminate against certain materials because "they are here, but not from here," feels more like the exclusionary rhetoric of a white nationalist than the perspective of any critically reflective creative practitioner. These people wish to keep resource streams pure, and therefore separate from the contaminating effects of those foreign, nonnative materials; this is absurd. The whole point of resource efficiency is to redesign sociotechnical systems so that we might more fully use the resources we have at our fingertips, here and now, regardless of their geographic origins or what far-flung corners of the world they originally came from. As you read this, you might feel irritated by my words. Maybe you are a designer-maker working with native materials and processes—a weaver, for example, producing textiles from locally produced yarns, dyed with pigments extracted from native plants and indigenous vegetation. This preservationist mind-set is an understandable response to the globalization of matter and the centralization of production, and all that it entails. These are vital practices, but we must also be wary of inbuilt prejudice lurking within our normative assessments of what is local and what is not.

We frequently see this type of binary opposition in the way food is described. Potatoes, for example, are not native to the United Kingdom. As a Brit who practically grew up on the things, I was rather surprised to have learned this only recently. The potato was originally domesticated by indigenous peoples of the Americas and then brought to Europe in the second half of the sixteenth century by Spanish colonists as a somewhat exotic foreign food packed with carbohydrates. Today we have many varieties of potato that we affectionately label as "British," "local," and even "native," yet most of these originated in the lowlands of southern Chile. So these humble tubers were not always "of this place," but at some point, they became indigenous as we accepted them as our own. In the age of the Anthropocene, polymers and precious metals are also becoming indigenous, whether we like it or not. They are here now and will continue to be here for centuries to come. Only a fool would overlook such a rich and bountiful material on the grounds of its foreignness.

The circular economy does not distinguish between material that is "of this place" or otherwise. Rather, the emphasis here is on keeping material in motion; whether that material has been here for millennia or minutes is of no consequence. In this way, resources are dynamic. They are matter in constant flux, flowing in and out of space. In the urban mines of tomorrow's circular economy, gold will be extracted from old computers, not ore; cotton will be harvested from worn shirts, not fields; and cobalt will be processed from unwanted flat screens, not a million tons of mud. In Anna Tsing's *The Mushroom at the End of the World: On the Possibility of Life in Capitalist Ruins*,[18] we experience the complex yet ongoing relationship between impending ruin, on the one hand, and the emergence of immense value, on the other. Tsing, a professor of anthropology at the University of California, brings our attention to the entangled nature of destruction and creation, using the extraordinary matsutake—one of the most valuable mushrooms in the world. It grows only in human-disturbed forests across the Northern Hemisphere. Strangely, the matsutake only seems to appear where human destruction has occurred and the ground has been tainted in some way. At a time of massive human destruction and growing fear of the Anthropocene and its implications for the future of life on earth, Tsing's work teaches us how adverse conditions can lead to new forms of value that could not have emerged any other way.

Yet while urban mines may be localized in the form of unwanted products, our current industrial means of disassembly, processing, and manufacture remain globally diffuse. These centralized processes of industrial production were never designed to deal with the new material reality we find ourselves in. Indeed, thinking of products as "material stores" changes the way we think about the design of products themselves, in addition to boldly reimagining entire sociotechnical systems through which production and consumption continually flow.

We live in a predominately linear economy and have largely done so for the past century. This linear economy was born out of a preindustrial mindset, in which the planet's resources were considered limitless and infinite. Our industrialized model of production is based on a linear, one-directional flow of resources and value. Essentially, our linear economy may be characterized by a straight line, with inbuilt social and environmental destruction at either end. The circular economy grabs both ends of this line—the beginnings of products, and the ends of products—and bends them round to form a circle.

The transformative effects of this design-led system-level change are significant and will be decisive in determining the resource security of future generations. They will also bring about a complete overhaul of the short-sighted way we perceive underused products as either "problematic waste" or "rich material stores." As Duncan Baker-Brown, an architect and advocate of the circular economy, describes, "There is no such thing as waste; just stuff in the wrong place."[19] Indeed, traditional linear models of production and consumption—often referred to as "cradle-to-grave"—have proved to be disastrous, especially from an environmental point of view.[20] The Ellen MacArthur Foundation describes clearly how new types of economies must emerge, aimed at transitioning the paradigm from a linear model to a circular one.[21] Similarly, Joanna Boehnert, author of *Design, Ecology, Politics*, forcefully states: "Human impacts on planetary processes in the Anthropocene require new types of ecologically engaged design and economics if the necessary technological, social, and political transitions are to take place."[22]

While the simplicity of the metaphor—a circle—is helpful in some ways, it can also be misleading. The circle wrongly implies a returning of material to the point of origin. This makes sense only at the level of the biosphere, not at the level of products, materials, or complex and geographically diffuse supply chains. This distinction is rarely made and represents a common source of misinterpretation by designers looking to engage in this relatively new arena. Of course, the term "circular economy" provides a useful handhold for us to collectively grasp; this is why I use the term here, for example. The caveat is that materials need not always return to their point of origin, as the metaphor of a circle implies. To some readers, this may seem obvious, but I see too many "circular design" proposals, which are essentially glorified product take-back schemes. This is one approach, but the potential of the circular economy is greater than this.

The notion of a "mesh" may be more useful. In a mesh, materials flow freely from one junction to the next. The lines of the mesh represent material pathways; the knots or intersections represent assemblages. Or as Timothy Morton puts it: "The whole is a mesh, a very curious, radically open form without centre or edge."[23] What is helpful about the mesh is that it possesses a level of intricacy that the circle does not. Think of a large sheet of chicken wire with its hexagonal negative spaces, defined by steel wire. Now, scrunch up that mesh into a ball to form an intricate, three-dimensional mesh. The term that perhaps best describes this form of aggregation is "meshwork," which the social anthropologist Tim Ingold uses to describe a temporary "entanglement of lines" that, unlike the lines in a network, "do not always connect."[24] According to Ingold, "These lines are independent and connect in knots, which evoke movement and growth but, significantly, do not prevent other lines from overtaking them; rather, the lines keep going like loose threads," a material system in which matter flows in many directions and can travel backward and forward along myriad defined paths.[25]

Despite the limitations of the metaphor, "circular design" encourages us to rethink business models and how we make products, and to consider the interconnected systems surrounding them. But we also need to think about the materials we use. Indeed, not all materials we use today are fit for a circular economy. Some contain chemicals that are hazardous for humans or the environment. Additives are often unintentionally used for performance reasons—such as coloring, flexibility, and durability—but ways exist to design them out. If you can choose materials that are safe and circular, you can build a better offering for users while ensuring that the products and services you create fit within a circular economy.[26] Reframing the systemic conditions of production and consumption as circular, not linear, opens significant opportunities for sustainable change. New ways of designing, making, and using become of commercial interest, and seemingly immovable socioeconomic barriers to progress crumble away. These new ways are exemplified by innovative products like Futurecraft Loop from Adidas—a monomaterial performance running shoe that can be returned to Adidas at the end of its useful life and ground down to make more shoes. This ongoing process of material reincarnation allows the product to have many lives, not just one.

In the circular economy, value may be created at several points throughout the product's lifetime, not just at the point of sale. For example, "Sharing platforms maximize the use of underused assets and enable companies to create additional value from products that have already been sold; leasing arrangements can increase margins, enable new revenue streams, and make a product more accessible for people; material recovery and recycling leverages technological innovations and capabilities to recover and reuse waste to turn it into resources; and, designing a second life into products enables companies to sell the same product again and again, or to offer new repair, upgrade and maintenance services."[27] Through these kinds of design-led approaches to value creation, businesses are incentivized to extend product life. Operating from a small town in Wales, Hiut Denim Company makes jeans to last, offering leasing models and free repairs for life. Hiut encourages customers to join the "No Wash Club" and not wash their jeans for at least six months. Every pair also comes with a unique "history tag," encouraging wearers to engage in storytelling through social media.[28] By increasing the product's life cycle, you correspondingly increase opportunities for value creation. Indeed, designing for a circular economy is not straightforward. "Gone are the days of 'sustainable' or 'eco' design, when a simple change of material to a recycled alternative would give a project environmental credibility."[29] Now, to understand all the interconnected facets of the problem, we must expand our thinking to include the whole system, from resource extraction and manufacture to use and disposal. From an economic perspective, circular supply chains also guarantee a steady and predictable flow of materials. This allows materials to be used repeatedly while insulating brands from the volatility of resource markets and their knock-on fluctuations in raw material prices. When businesses transition to a circular economy, they take control over their material flows. Whether during production, use, or end-of-life, businesses become custodians of material. Through this process, new ways of designing emerge that reimagine products and services as being suspended in a continual state of flow.

Used but Not Owned

We are all familiar with the notion of "ownership." Indeed, if you have made it this far in the book, you will likely hold some degree of expertise on the matter. Conversely, "usership" is a less familiar concept. In part, this is because "usership" is a neologism, in the process of entering common use, and has not yet been fully accepted into mainstream language. Yet despite the term being unfamiliar to most people, it points to a behavioral phenomenon that each of us is intimately familiar with. Think of the word "usership" as you might think of the word "relationship." Let us try a quick thought experiment. Think about someone you have a relationship with—a friend, a family member, or a partner. It could be anyone, provided you consider yourself to have some form of relationship with the person. Now, are you with that person right now? My guess is that you are not. Even though you are not physically with that person at this moment, you are still in a relationship with him or her. The relationship does not end when that person walks out the room, and restart when he or she comes back in. It continues through the absences when you are not directly interacting with each other. Usership is like this: a continually unfolding process of engagement, punctuated by moments of direct interaction that are few and far between. These moments of direct interaction—like touching a thing, pressing a thing, or swiping a thing—tend to be what user experience designers primarily focus on. But, in terms of usership as an expanded, longitudinally rationalized process of use, these moments are all too fleeting to be the sole focus of design attention or the basis of any universal principle of subject-object interaction. Instead it is more useful to think of usership as a discrete state, something to be considered and designed for. Indeed, usership is an active condition involving serial "engagements" (a term Merleau-Ponty uses to describe the act of "coming to know our world") with designed things.[30] This expanded view of person-world interaction more accurately points to the lived experience of being a user, engaging with a dynamic assemblage of things over time.

Considering the longitudinal, entangled nature of usership expands our understanding of the state of use and its temporally extended nature. It begins to lift us out of the reductionist, intellectual nosedive that "human-centered design" has dragged us into. It invites us instead to reconceptualize the process of use as a continuum, a perpetually unfolding relational state that flows, whether you are physically interacting with the product or not. On the most superficial level, "use" happens during moments of direct interaction between the subject and the object. This kind of clumsy thinking defines users as though they were light bulbs with on/off states; either they are using the product, or they are not. Despite being the most common denotation of "use," it is pure folly. Although easy to conceptualize, such simplistic accounts of subject-object interaction are deeply unhelpful and offer limited value to design. Understanding interaction on such a shallow level sorely overlooks the governing influence of the meanings, contexts, and systems in which these complex interactions between people and the designed world unfold. Sure, in starkly lit user testing labs, one may be fooled into believing that this cartoonlike reality exists, and "use" happens in a vacuum, undisturbed by contextual externalities like mood, weather, time, hunger, or illness. Yet beyond the predictable and controlled conditions of the lab, out in the wild, we find no such well-ordered contexts of use; they simply do not exist. Use happens within dynamic and relational systems, and these systems are constantly changing. They are also nested within other systems, which themselves are dynamic and shifting. Through their interrelation with one another, they shape our experience of the world. User experience, therefore, is an emergent property of the systemic conditions in which the process of interaction unfolds. Indeed, the first step to solving an intractable social problem like overconsumption is to understand the multiple systems in which that problem simultaneously resides. If you do not, you might find yourself investing in a solution that is ineffective, takes more time or resources to implement, or even makes a problem worse.[31]

On a cognitive level, we are engaged with designed experiences all the time. Frequently, these interactions occur unconsciously, when the product is not even present at hand. Interaction, therefore, is a far more expansive territory than is commonly assumed, and is certainly not just what happens when we click, swipe, or press a given thing. In this way, "Common notions of interaction, those we use every day in describing user experience and design activities, are inadequate."[32] Rather, interaction is an omnipresent condition, which occurs at different levels of scale—products, environments, services, and systems—from the electronics in our pockets or the built environments we move through, to the news feeds we subscribe to and the oppressive political systems we push hard against. On an experiential level, we use all these things, continuously. There certainly are different levels and depths of use, and if we can become clearer and more familiar with these levels and depths, we can design more appropriately for them. Interaction design, then, is a way of framing the ongoing relationships between people and the worlds they inhabit, a way of framing the intent of all design. As service designer Hugh Dubberly describes, "Every designed object offers the possibility for interaction, and therefore, all design activities can be viewed as design for interactions. The same is true not only of objects but also of spaces, messages, and systems."[33] At the level of human experience, therefore, all design is interaction design.

Of course, not everything has to be owned to be used. On the contrary, many of our day-to-day interactions with the designed world involve things we do not own. This morning, for example, I stood at a bus stop in downtown Pittsburgh, shielded from the rain by the limbs of a vast maple tree. I did not own any of it—not the bus stop, the pavement, the tree, or the rain. Yet the experience itself was "mine." Paddle in Lake Michigan, snooze in Central Park, or camp in the Adirondacks; none of these environments—or the designed and nondesigned things within them—are owned by us. Yet our interactions with them are meaningful, and the experiences we have in these environments do belong to us. In some ways, our life experiences are shaped through encounters with places and things not owned by us. These kinds of "open-access regimes" surround us and range in scale from wonky paving slabs and dry autumn leaves to climatic conditions and the stars in the night sky.[34] One might even argue that some of our most meaningful interactions with the world have this open-access character. Of course, these experiences are owned in a civic capacity; they are just not owned exclusively by you. A city, for example, is a complex but incomplete system. In this mix lies the capacity of cities across histories and geographies to outlive far more powerful but fully formalized systems—from large corporations to national governments.[35] And so we are permitted access to these complex, incomplete, and continually unfolding places, but our access is conditional. For example, I can snooze under a tree in Central Park, just not for too long before being told to move along by park security. I can paddle in Lake Michigan, but I must not be naked. I can freely camp in the Adirondacks, just not for more than three nights without a permit. These are all reasonable constraints set in place to regulate against the irresponsible use of a shared resource and avoid a "tragedy of the commons,"[36] to use the British economics writer William Forster Lloyd's term. The tragedy he refers to indicates a state in which individuals behave according to their own self-interest, at the expense of the common good. Forster Lloyd's seminal concept is frequently misattributed to the American ecologist Garrett Hardin, whose 1968 article built on and popularized Forster Lloyd's work almost a century later. Hardin warns of how the underlying logic of the commons generates tragedy. That "ruinous destruction is the destination toward which all people rush, each pursuing their own best interest in a society that believes in the freedom of the commons," more than anything else.[37]

The destructive phenomenon of self over community is not just found in physical spaces like national parks or urban spaces; it also manifests in the digital world. For example, social media platforms like Facebook, YouTube, Instagram, and Twitter each provide a kind of "public space" or "commons" that we are all invited to use, and misuse. Of course, when the digital service is free, you are the product, as it is the resource that uses you (e.g., data harvesting, analytics, rich user profiling), just as much as you are using it (e.g., posting, learning, bragging). Or as Crawford and Joler put it: "The user is simultaneously a consumer, a resource, a worker, and a product."[38] Digital systems that are openly shared by entire communities also tend to get corrupted in the end. While these forms of open system are enjoyed and respected by the majority, they attract their share of bad actors as well: people and organizations who exploit free resources for money or perverted motives.[39] That YouTube, for example, is open and free allows all kinds of creativity to flourish in ways that are not enabled by the gated nature of the entertainment industry. The tragedy is that this openness also empowers pornographers and propagandists for terror. Even in the digital commons, the tragedy reliably plays out, just as it always has—whether it be angry users shouting to gain attention, or corrupt governments propagating fake news to shape public opinion and rig elections. These are the denuders of the digital commons, overgrazers using the resource in a way that meets their own needs at the expense of others.[40]

Using Together

In most capitalist societies, the compulsion toward absolute ownership persists. It appears we will always have a need for things that we own completely, things that are ours and ours alone. Ordinarily, people are reluctant to share their cherished, "high-stakes" possessions with others. The meaning of these kinds of objects is deeply personal, and thus such things are considered irreplaceable, and certainly too precious to place in the careless hands of another. In the act of sharing such objects, therefore, the risk feels particularly high. However, these are minority objects and represent a relatively small portion of our material worlds. Much of our stuff is just that, "stuff," multitudinous things to which we have little or no emotional attachment or positive meaningful association. In fact, many of the things we own, we do not even like anymore. We simply do not care enough about those items to replace them with ones we do care about; they are useful, but that is about it. In this way, many of our low-stakes possessions could quite readily be shared. Low-stakes products might typically include utilitarian items like hedge trimmers, corkscrews, bicycle pumps, and ladders. These kinds of products make up a large portion of our materials worlds. They all lie ready for action but are seldom called on; highly useful, underused things to which we have little or no meaningful connection are ideal for sharing.

Sharing our things takes us toward more "we-based," not "me-based," forms of material culture. It represents a bold move in the transition toward a society that gets by with fewer things. Yet sharing faces many experiential barriers. As mentioned earlier, sharing can be inhibited by our tendency toward possessive individualism, in which individuals are conceived as the sole proprietors of their skills and owe nothing to society for them.[41] Sharing motives are also held back by the "endowment effect." As the cognitive neuroscientist Christian Jarrett explains, we form connections to things once owned, and value items much more highly when we own them. Jarrett refers to the Hadza people, an indigenous ethnic group in north central Tanzania. Isolated from modern culture, the Hadza do not exhibit the endowment effect. Instead they live in an egalitarian society where everything is shared and the notion of ownership is a collective one.[42] To the Hadza, all things are "ours," and never "mine." Indeed, sharing can be understood as a social practice we so-called moderns have only recently been cultured out of.

Weston Baxter speaks of "contamination interaction" and the way in which shared objects may create feelings that they are foreign or contaminated and belong to someone else. This could occur through hygienic contamination (threatens feelings of hygienic cleanliness), utility contamination (functional attributes are damaged), or territorial contamination (object seems to be "marked" by another person). Collectively, these forms of contamination deter people from sharing their possessions with others, while also deterring people from using the possessions of others. In part, a pathological aversion to contamination drives our reluctance to sharing, not just selfishness, greed, or covetousness, as is so often assumed. Some level of sharing still goes on, of course. Even in societies awash with the swill of free-market capitalism, we continue to engage in the sharing of "tangible" and "intangible" assets, albeit at a significantly reduced level. Tangible assets include things like power tools, cars, and clothes. These are physical things that one person has, but another person requires access to for a specified period. In contrast, intangible assets might include things like technical skills, free time, and physical strength. These are "things" a person "has," and another person requires access to. These assets are simply nonphysical.

To many, the model of "sharing as a means to access x" closely aligns with the common social practice of "borrowing." With borrowing, the asset is still yours, but others are trusted to use it for a while. Think of the last time you lent someone a book or a coat. Borrowing is "keeping" while also "giving," a concept explored in the American anthropologist Annette Weiner's seminal text *Inalienable Possessions*.[43] She investigates the category of possessions that must not be given or, if they are circulated, must return finally to the giver (e.g., books, clothing, tools). The paradox of keeping while also giving points to the indebtedness one feels when being allowed to use another's possessions. This indebtedness itself provides a form of social currency, "owed" by the borrower to the lender. Underpinning this indebtedness is a complex dynamic of trust, which must be honored and respected. As Belk points out, whereas much economic activity avoids feelings of commitment, sharing and borrowing promote it, with the potential for lingering indebtedness and residual feelings of friendship.[44] In this sense, the design of effective sharing experiences can elevate the meaning and value of our interactions with the material world, by creating opportunities for us to emotionally connect with one another: a temporary melting of the boundaries separating yours from mine, and you from me.

Brian Burns, a professor of industrial design, describes how our materialist obsession with the "means" has disconnected us from the "ends." For example, the means (e.g., toasters) are the primary focus of our attention, relegating the ends (e.g., toast) to a mere consequence of the means. In these instances, toast is seen as a by-product of the toaster, rather than the primary object of our attention—such is our enthrallment with material goods. In this means-obsessed modernity, we think cars, not mobility; smartphones, not communication. Louise Downe, director of design for the government of the United Kingdom, describes an emerging situation in the field of service design, where the perception of a service is shifting unhelpfully away from "service as process" toward "service as thing." To this effect, she describes how good services are verbs, and bad services are nouns.[45] In other words, services are dynamic processes with an emphasis on what you do (verbs), not fixed and static things known for what they are (nouns). Thinking of services as verbs emphasizes their dynamic properties, in turn exposing the relatively static qualities of most designed products. Of course, products are often the only tangible element of a service, the only part we see, touch, and hold. It is understandable, then, that we might tend to place greater emphasis on the tangible elements of a system that have been materialized, while remaining relatively ignorant of the rest.

The societal shift from ownership to usership can lead to more environmentally responsible behaviors and is about far more than simply buying less stuff. With cars, for example, when people's mobility costs shift from being fixed (owning your own car) to variable (renting access to a pooled car), people make more efficient decisions about when they need to drive. Indeed, studies have shown that the average car sharer drives 40 percent less than the average car owner.[46] Businesses play a key role in this, and many "sharing economy" start-ups are emerging in this space. At its core, "the sharing economy is about the sharing of idle assets, usually via tech platforms, in ways that produce economic, environmental, social and practical benefits."[47] Or at least it is supposed to be. Sadly, the "sharing economy" is a term frequently slapped onto ideas that promise a means of matching supply with demand, but zero sharing and collaboration are involved. At its worst, the sharing economy has shown itself to be overwhelmingly an antiregulatory, precariat-creating way of monetizing social interactions.[48] Commercial activities are dressed up as sharing—a kind of "sharewashing" in which the language of sharing is used to promote new modes of selling.[49] This, regrettably, offers nothing more than a depressingly logical conclusion to the merciless endeavors of profit-driven, hegemonic corporations. For example, Airbnb is not sharing. It is a platform to support the monetization of underused private space. Airbnb is as much about the sharing of rooms as the average hotel is about the sharing of its rooms: "Yes, you can share our rooms; prices start from $99 per night." Sharp distinctions persist between these models of business-to-consumer sharing practices and the kinds of peer-to-peer practices one might witness at a community level. Sharing has huge social potential, but regrettably, this potential has been obscured by the bastardizing practices of a handful of corporate racketeers, probing the market in search of ever more cynical means of making money.

Despite the repertoire of bad examples—and it is a rich repertoire—it makes sense to share. From an ecological perspective, the case is clear. Sharing allows us to do more with less, maintaining high levels of human activity while dramatically reducing our pressure on resources—provided, that is, that these things are designed for longevity and ease of repair. Sharing also brings significant cost benefits to consumers. When you have ready access to the things you need, when you need them, you no longer need to buy and maintain one of everything. From a social perspective, sharing also provides a means to engage and connect with one another. The deepest social value lies in peer-to-peer sharing. For example, Ann Light and Clodagh Miskelly, researchers of interaction design for social change, show how locally developed initiatives can differ in tone, scale, ambition, and practice from those drawing the lion's share of attention in the "sharing economy" debate. In locally situated cases, they observe an emphasis on organizing together to create shared spaces for the collaborative use of resources and the joint ownership of projects and places, whereas many large-scale global models of sharing feature significant elements of renting, leasing, and hiring and focus only on resource management, sometimes at the expense of community growth.[50]

Critique of the sharing economy is well established. Today the capitalist motivations underpinning this phrase—and those who peddle it—are self-evident. The term is itself unhelpful in the way it situates sharing within a primarily economic context, very different from the postcapitalist mode of material engagement that design-led systems of shared use initially sought to enable. While sharing might contain an economic dimension, its value as a social practice is more far-reaching. What we should be transitioning toward is a "sharing culture," not a "sharing economy": one that includes the economic but is not dominated by it; one that is far broader in its social and political scope.[51] A growing array of peer-to-peer sharing platforms support direct engagement between people, at both global and local levels, whether sharing digital stuff across the world on file-sharing platforms LimeWire or BitTorrent, or sharing physical stuff with your neighbors with the help of sites like Peerby or Freecycle. Such platforms help us think beyond oxymoronic tropes like "sharing economy" to instead focus on "sharing culture" and how we might design more optimal conditions to enable a shift to collective, locally situated ways of making, using, and repairing our things.

Making Together

We have become divorced from the process of making. Today we find ourselves exiled, cut out, and banished from this vital aspect of our creative selves. In a globalized world, corporations are the makers; it is they who hold the tools, the materials, and the means of production. In this highly centralized world of mass production, users are not welcome; they would only slow the machines down. Yet with this disconnection from making comes a requisite disconnection from one another. Making is a social practice. It connects us to the physical world and its dynamic properties, and to the social world and its dynamic properties. Today we find ourselves locked out of these vital creative and social processes. The world we now inhabit is largely out of our hands. Now, instead of making our futures, we simply consume them. Let others design and make your things, then choose and operate the things you like; this is the dominant mode of engagement with the material world. The professionalization of making has come to dominate our experience of everyday material life to the extent that we now feel entirely detached from even the simplest forms of making. For many, making is not a part of their consciousness and represents nothing more than an inconvenience. We have become so experientially divorced from the practice of making that if you were to make your own dress, or even fix your own shoes, you would be considered something of a rebel, refusing to fall in line with the passive, emotionally disconnected horde.

Giles Smith and Amica Dall of the London-based architecture collective Assemble use the contemporary craft movement to look at our changing relationship with labor, value, and the material world. In a radio program titled *The Sympathy of Things*,[52] they explain how since the Industrial Revolution (ca. 1760–1830) and the introduction of mechanized production processes, economies of scale have governed our lives. Smith explains how "if you want to get into making teapots, for example, you make the teapot making machine once; then you can make as many teapots as you like, and the more you make, the less each teapot costs"—provided, however, that the teapot is one type or pattern and there is no variation from one teapot to the next.[53] Such is the principle of mass production, and designers educated within the centuries-old mind-set of mass production will feel this is second nature—a stating of the obvious. Yet these established norms also hold us back in our thinking. For example, in digital manufacturing, these norms no longer apply. Dall uses the example of the laser printer to explain this. She describes how "on a laser jet printer, it does not matter if you print the same page of text 100 times, or if you print 100 different pages—it costs the same and takes the same amount of time."[54] Digital manufacturing works in similar ways. It reopens our capacity to make what is needed, when it is needed, and where it is needed. If you stay within the broad parameters of the machine, you do not have to make identical copies and can change the design of a product as many times as you like, without slowing production or raising cost. This allows for a degree of idiosyncrasy to mass production while also creating opportunities for people to participate actively in the creation of things, and not just passively in their consumption. According to Smith and Dall, "This form of 'mass customization' enables a tremendous increase in variety without a corresponding increase in costs; allowing for the mass production of individually customized goods and services."[55]

Much of this newfound agility in manufacturing is enabled through digital processes, ranging from "additive processes" like stereolithography or 3D knitting to "subtractive processes" like water-jet cutting or CNC milling. SPACE10, a Danish design lab exploring more sustainable ways of living, describes how factories of the past cranked out high-volume identical products and shipped them around the world. With powerful, more accessible industrial-grade machines, such as 3D printers, CNC machines, and laser cutters, the process of production is turned on its head. Instead of shipping container loads of identical products, we can send digital files and have things made locally, wherever and whenever they are needed, with materials that can be sourced, recycled, or grown locally and sustainably.[56] Decentralized processes such as these take us toward more flexible, open, and lean forms of codesign and manufacture that can adapt and change to meet individual customer needs and requirements while continuing to operate at scale.

A good example would be Unmade, a company that designs software to enable consumers to customize their products before manufacture, making them highly personal and tailored to the individual. Unmade uses a process it calls "curated customization," a partial opening of the design process that invites users to participate. The result is a product that feels coproduced—a creative collaboration between the user and the brand. In this scenario, Unmade maintains control over the majority of design decisions, but the customer feels more engaged in the experience of creating a product.[57] Unmade's list of partners is growing and includes Selfridges, Opening Ceremony, MoMA, and Christopher Raeburn, who used the software to enable customers to "edit" key pieces from his spring–summer 2016 collection. Pieces typically sell for around £60 to £200 and have been deliberately designed to have wide appeal. In 2019 Unmade developed a new collaboration with New Balance. The partnership combines New Balance's rich manufacturing heritage with Unmade's state-of-the-art digital technology and allows customers to create their own knitted shoe upper with a choice of graphics, color, and text. The result is an entirely bespoke product, manufactured on demand.

Digital additive manufacturing processes like 3D knitting involve no wasted materials, because the machine makes the pieces to the exact size and requirements needed. Such technologies also support local production while enabling an on-demand production model in which only what is needed is made. In the fashion sector, this is particularly important. Right now, the fashion industry operates speculatively, meaning that approximately 20 percent of everything it produces fails to sell and goes to landfill or is incinerated. Of course, questions surround our curiosity for one-off, unique items. Is this what people want? One might argue that a certain comfort resides in uniformity, and the idea of a fashion "line" has an appeal of its own. We might also suggest that mass customization introduces an element of risk. After all, most people are not fashion designers and, if they are anything like me, can barely put an outfit together, never mind customize a piece of bespoke knitwear. Unmade deals with this cleverly by allowing customization to take place, but within a clear set of well-defined constraints. This affords users "just enough scope" to incorporate a bit of themselves into the piece, without departing too far from the original design form. In 2000, Nike launched NikeID, a service that allowed shoppers to customize shoes with the color, fit, and feel that reflected their own style. The options were initially limited to only one type of shoe but still offered customers an introduction to the experience of product customization. Today Nike's customization platform, known as Nike by You, offers users several levels of engagement across a range of shoe categories: gym, football, running, baseball, and so on. By offering a wide range of customization options, products in this rapidly expanding range spark richer forms of dialogue between designers and consumers and begin to reconnect everyday people with the process of creation.

As a social process, making matters. It provides a means through which people can communicate something about themselves, through a thing they have made, which can then be seen and experienced by others. This reflexive social capacity is present in practically all made things—whether a hat you knitted for yourself or a website you coded from the ground up. In *Making Is Connecting*, the British sociologist and media theorist David Gauntlett describes the importance of people being able to make things, express themselves, and share ideas through the things they have made in physical and digital spaces—whether a jeweler or a woodworker, a zany YouTuber or a crack coder.[58] The scale of the making is not important; that you are engaged in the process is. Gauntlett's work explores everyday acts of making and sharing, expressed through both analog and digital means. He argues that through making things, people engage more intimately with the world and create meaningful connections with each other; making has a human capacity that connects us in profound and complex ways. Making things, and then sharing what you have made with others, is now easier than ever before, thanks to the connective capabilities of social media. Due to the image-led capacity of Instagram and Facebook, for example, a tray of brownies emerging from the oven, or a salvaged rocking chair on your workbench awaiting the final coat of magenta paint, has the potential to be photographed, shared, and liked. One might even argue that the reward gleaned from this sharing and liking provides a powerful motivational driver for making activity.

Lack of access to equipment, space, and expertise presents one major barrier to making. Today, few people possess the tools or know-how to make or repair even the simplest of things. Maker spaces aim to offer a solution to this. We are seeing a rapid growth in making activity at the community level. Over the past decade, maker spaces have bloomed in worldwide popularity. More than two thousand such spaces are active today, with many more on the way. Maker spaces are places for creating, collaborating, learning, and sharing, using high-tech to no-tech tools. These spaces are open to kids, adults, and entrepreneurs and have a variety of equipment, from lo-tech options like hammers, saws, drills, soldering irons, and sewing machines to hi-tech making equipment like 3D printers, laser cutters, Arduino kits, and CNC vinyl cutters. The wrong way to think about fab labs is simply access to machinery, a way to get your hands on the tools you need to make the stuff you want. Although this aspect of fab labs exists, we might more accurately understand machines as tools to empower processes. In this way, when you enable people by giving them technologies, people develop agency and become able to solve their own problems. In *Futurekind: Design by and for the People*, Robert Phillips describes the way citizen-led forms of creative practice unify expertise and communities and enable people to actualize ideas together, finding their own solutions to their own problems and circumstances.[59] Phillips is a senior tutor in design products at the RCA and a fierce advocate for public participation in socially and environmentally responsible design. He describes how in "bringing together digital communication and engaged online communities, products can be designed by anyone for social and environmental good."[60] In this way, maker spaces form a critical component in the shift from ownership to usership. A well-equipped community maker space has the tools and shared expertise to enable people to make, repair, and modify things together at a local level. In this way, the decentralizing effects of maker spaces underscore their contribution to the transition toward a society that does more with less. The Fab City Global Initiative takes this a step further, through an expanding network of locally productive, globally connected cities. At the heart of this network of cities, regions, and countries is a pledge to work toward producing everything they consume by 2054. In the Fab City, people need things and then make the things they need. It is a platform that enables and speeds up social innovation processes while proposing a radical shift in the way we make and use things.

Of course, the jury is still out on the environmental benefits and social efficacies of maker spaces as an alternative to the massive, centralized systems of production born out of the Industrial Revolution. The fields of science and technology studies, and history of technology and design, remain particularly active in their ongoing critique of community-based maker spaces and the actual difference they make. This critique is important, as it invites us to consider the limitations of these relatively new approaches and also serves to temper the overly enthusiastic rhetoric that tends to come with access to digital fabrication processes. In *The Charisma Machine*,[61] Morgan Ames argues that it is a fairly utopian and outmoded vision that motivates the initiation of projects trying to use technology to positively disrupt social and cultural development. Despite grand claims, these initiatives frequently rest on a flawed technocentric vision that assumes technology will always play a central role in any human development context. Her book offers a cautionary tale against the allure of technology hype and the problems that result when utopian dreams drive technology development.

We should not be making all our things ourselves, as Thomas Thwaites explores in his book *The Toaster Project*. Thwaites is a designer whose polemical work helps us speculate on—and think more purposefully about—the knotty problems riddled throughout our sociotechnical systems. He describes his quest to build a simple electric appliance from scratch—a significant challenge, given that a typical low-cost toaster contains more than four hundred individual component parts and uses about a hundred different materials. On this noble quest, Thwaites mined ore for steel and derived plastic from crude oil to experience the myriad industrial processes embedded in even the simplest everyday product. He succeeded in the end and was able to produce one working toaster, although it took him nine months to make and cost about 250 times more than the cheap toaster he bought in a store at the outset of his project. Importantly, it is not that Thwaites's toaster is expensive; it is that other mass-produced toasters are misleadingly cheap. "It takes an entire civilization to build a toaster," says Thwaites, whose book might well be a cautionary tale for an interconnected society, born into excessively centralized models of production.[62] It is through this centralization that we have become alienated from the essentially human capacity to make things. Today, our once-native skills and understandings of how the physical world works have atrophied through disuse. This former capacity, once critical to survival, now lies dormant within us like a "vestigial structure" with no purpose in the current form of our species, like the coccyx or wisdom teeth.

Not only is there a social side to making; there is a social side to mending also. The Restart Project, for example, is a social enterprise that encourages and empowers people to use their electronics longer to reduce waste. Restart was founded in 2013 out of a deep frustration with the throwaway, consumerist model of electronics that we have been sold, and the growing mountain of e-waste that we are leaving behind. Restart's mission is to help people learn how to repair their broken electronics and rethink how they consume them in the first place.[63] One of the company's main points of intervention within the system is to help organize "restart parties," where people teach each other how to repair their broken devices, from tablets to toasters, a volunteer-run and truly people-powered enterprise that aims to fix our relationship with electronics. Through these collaborative mending practices, meaningless rubbish is converted into meaningful stuff. New stories are coproduced by participants and woven into the experiential layers of everyday possessions through the practice of repair.

In the context of usership, fixing is a creative and social act, not just a necessary one. While we can fix things alone, fixing is better shared. As people share skills, advice, and tools, these gatherings become a site for social exchange, a focal point around which large numbers of previously disconnected individuals might gather and, for a handful of hours, feel connected to one another. This approach offers a sharp contrast to the idea of the "repair shop," which outsources fixing to a centralized, anonymous community of "experts." In this subordinate model, repair is just another way to make money; we are kept dumb, isolated from one another, lacking in agency. Socializing making and repair enables two levels of transformation: making and fixing broken products, and making and fixing broken social systems. Indeed, using together, and making together, provides a vital context for collaboration; products are made and repaired, just as communities themselves might be made and repaired. Through these practices, we see sharp reductions in consumption and waste and a proliferation of social cohesion and localized positive action.

7 Design That Lasts

Optimal Conditions for Change

Design's historic failure to acknowledge the wider systemic conditions in which consumption, use, and disposal take place represents one of the deepest points of our profession's negligence. The design challenge is to ensure that products last in the hands of their users, and not in the depths of a landfill site. This is a subtle distinction, but an important one. We are often duped by claims of technical robustness, or material strength, and take this as a guarantee of product longevity. Indeed, there is no point in designing physically durable products if users lack the desire to keep them. When we do this, we are simply designing highly resilient forms of refuse for future generations to contend with. Of course, physical resilience has a part to play in the design of things that last, but physical durability alone is not enough. As I discussed at the outset of the book, products already last; they always have. Just because a given product is no longer wanted, it still "lasts." Today, our world suffocates beneath an ever-thickening crust of discarded stuff, much of which still functions perfectly in a utilitarian sense. Atop this crust of orphaned products, factories churn out even more stuff, which, after a brief period of use, will meet the same fate as its subterranean counterparts. These ecologically blind practices have contributed to the formation of Earth's current geologic time period, the Anthropocene, an anthropogenic era in which geologic, hydrologic, biospheric, and other Earth system dynamics have been irreversibly changed by humans. The Anthropocene is "a new period either after or within the Holocene, the current epoch, which began approximately 10,000 years ago with the end of the last glacial period."[1] Such is our impact on the planet that we have changed its millennia-old geological makeup and permanently altered the Earth's natural, systemic processes that for more than a billion years have provided the necessary conditions for life.

It is our connection to products that fails, not simply the product itself. The kind of waste this experiential breakdown generates is a symptom of relationship breakdown between the user and the product. Through this emotional failure between people and things, value and connection "break," leading to the discarding of the outgrown object by the transcendent other. Landfill sites are stuffed full of these kinds of "broken" things, which function as well as they did when new but have since failed in some other "experiential" way; they are not meaningful anymore. To be more specific, the character of this meaning has changed in a way that renders the product less desirable from the user's perspective. This is a deeply subjective process, but one that unfolds continuously in everyday life.

As in the stock market, so the value we perceive in our possessions fluctuates constantly. Most of the time, we are unaware of these ebbs and flows in the meaning and value of our things. Only occasionally do they reveal themselves to us. Even damaged or dysfunctional products live on, frequently enduring centuries of slow decomposition after a relatively fleeting use phase. If not recycled, a plastic stirring stick will last about two hundred years despite only being in active service (stirring) for two seconds or so. And so design already does last. But this "lasting" commonly happens in places and ways that were not part of the original design intent; their longevity was unintentionally designed. Ordinarily, the end of a product's life is barely considered at all, and so to describe it as "not part of the original design intent" is something of an understatement.

Our designerly preoccupation with onboarding experiences (how things start) is shadowed by a relative neglect of offboarding experiences (how things end). As an industry, we have become fixated on an outmoded model of user engagement, with its sole emphasis on beginnings. In a streamlined world, with consumers moving freely between multiple product relationships, clean and considered endings will be a competitive differentiator.[2] This, in many ways, is the greatest professional challenge we face in moving toward a culture of design that lasts: embracing the beginning, middle, and end of a product's journey. Widening the scope of design's concern pulls all these phases of the product's life into view. At this point, we acknowledge the entire life cycle of a product as wholly designable and far richer in its potential for driving social, ecological, and economic change.

I have discussed the negative impacts of short-lived products at length. Today, few would challenge the assertion that we in the global North inhabit a predominately throwaway society. While counterarguments or examples may suggest otherwise, the dominant mode of engagement with the material world remains a wasteful one. This consumption-oriented reality is characterized by relentless cycles of desire and destruction, through which we shop our way closer to an ever-evolving idea of "the good life." This future vision of how we would like to live is constantly shifting and inches ever farther away, just as we begin to close the gap. Despite our continued pursuit of this preferred state, it remains tantalizingly out of reach. This destructive behavioral pattern requires significant amounts of energy and resources to support and is fundamentally unsustainable. Over the past century, our insatiable consumption has degraded Earth's living and nonliving systems to an irreversible degree. Yet resource intensity has come to characterize everyday life, to the extent we barely even notice it, such is the normalizing character of consumptive behavior. Governments continue to confuse material throughput with progress, while individual citizens remain blindly attached to the belief that the more you have, the better off you are. We can think of this corrupted social condition as a form of mental pollution: the presence in the mind (environment) of something that has harmful effects (pollutant). Meaningful and lasting connections with each other, and our respective worlds, have been suppressed through this hedonistic culture of competition and individualism that conspicuous consumption has fostered over the past century. Indeed, so long as material throughout is upheld as a proxy for economic and social well-being, there is little hope.

In affluent parts of the world, most households already own at least one of each essential consumer product. Although it might not always feel like it, we already have everything we need, and have done so for some time. This has meant that the challenge for industry has been one of supporting people not in "having" but in "having more," by inducing dissatisfaction with what a person owns, to motivate replacement with a newer, better version. For decades, the capitalist system has operated on this basis, fumbling along to the cadence of this cyclical pattern of desire and destruction.

Against this rapidly cycling backdrop of consumption and waste, it sounds peculiar to be proposing a commercial environment in which we encourage people to keep products longer. Surely this will affect profitability or deplete market share? These are reasonable concerns, considering the capitalist roots of our economic status quo—a flawed model of how prosperity is to be reached that sees the level of unit sales as commensurate with the level of commercial success. Of course, it is easy to blame businesses for the sale of low-quality, short-lived products. But these ominous practices only persist because people continue to buy them. Indeed, designing products that last appears to "challenge many fundamental norms of economic policy, how businesses generate revenue, and people's relationship with possessions. This, however, appears on the verge of change."[3] Many businesses are beginning to explore design that lasts as a viable route to reducing waste and creating value.

The time has never been better to rethink the throwaway tendencies of modern design; the time to engage with design that lasts is right now. The conditions for design-led systems-level change will never be exactly right. If anything, the current state will always exist in opposition to the imagined future state, at least on some level. The systemic conditions from which the design-led change originates (the present) depend on the suboptimal reality we are trying to design our way out of. In other words, the current system likes it the way it is, and will defend against anything that does not resemble itself. It follows, therefore, that present conditions will always appear to present obstacles to new forms of design thinking that do anything other than reinforce the status quo. This will always be the case; such is the nature of systems change. As the economist John Maynard Keynes famously stated in 1936: "The difficulty lies not so much in developing new ideas as in escaping from old ones."[4]

Designing products that last is better for people and planet. As the sustainable design scholar Tim Cooper argues, "A societal shift toward increased product longevity will require greater understanding of people's experiences with their possessions: their attitudes to product lifetimes and expectations of them, their sense of attachment to possessions and how carefully they maintain them, and their decisions to dispose of them."[5] Cooper leads research in sustainable design and consumption at Nottingham Trent University and has been a strong proponent of designing products that last for several decades. He tells us there are many economic benefits from slowing cycles of resource use through designing products that last. Designing things that last supports us in creating new, meaningful relationships with ourselves, our communities, and our world. Furthermore, designing more "meaningful stuff" helps wean people off their relentless lust for the new and shapes new sustainable business models with the potential to transform the way customers engage with the world of goods.

Industry today is experiencing a seismic shift in thinking, transitioning from the design of short-life products based on a linear material system to longer-lived products and services founded on circular material systems. In 2019 Nike's chief design officer Jeff Hoke announced the company's commitment to "creating products that last longer and are designed with the end in mind."[6] This declaration from one of the world's biggest sports brands comes as a central part of their transition to the circular economy and to design that lasts. During the same year, IKEA began testing a range of subscription-based leasing offers in all thirty of its markets. According to IKEA's sustainability chief Pia Heidenmark, these trials give the company rare insights into the durability of its products, insights that feed back to designers and support them in making household goods that last longer and are easier to dismantle, repair, and move from place to place.[7] And so there is nothing niche about design that lasts. If anything, it is the guiding principle on which the future survival of brands will depend.

Longer-lasting products have the potential to present robust economic models for creating products, services, and brand-loyal customers—driving future sales, upgrades, service, and repair. This form of brand intimacy is supported by longer-lasting products, just as it can be eroded by products with excessively brief use careers. Most people want to buy products that last; many even make this criterion part of their buying decisions. A recent study by the UK charity WRAP (Waste and Resources Action Programme) showed that almost two-thirds of consumers felt it was more important that clothes were "made to last and look good for longer" than to be "fashionable."[8] It is a lazy, outdated argument to claim that shoppers do not care about product longevity. In fact, more than one-third of "British consumers report switching to a preferred brand for its environmental or social practices."[9] Take a pair of sneakers, for example. One thing we know about these sneakers is that someday, at some point in their life, they will wear out, and the user will need to replace them; that much we know. However, what we do not know is what brand the user will choose. Will a user replace them with a nice new pair from the same brand, or will the user be less loyal and try a pair from elsewhere next time? So often we hear opposition to longer-lasting products being based on such a story, where marketers express concern that allowing people to keep things for longer will lead to a drop in sales. Yet the inverse may well be true. By allowing customers to keep their products longer, you nurture a deeper relationship with both the product and the brand and, in so doing, increase the likelihood of stronger brand connections forming. Market share is secured and increased, and customers remain connected to brands in an ongoing relationship with a broader array of service touch points centered on repair, upgrade, and adaptation—not just one-off sales.

Designing products that last is a complex undertaking and requires us to embrace a multitude of external variables. These include a broad panorama of user conditions (e.g., beliefs, values, attitudes, habits); technical infrastructures (e.g., access to repair facilities, technical expertise, upgrade support); and contextual factors (e.g., sociocultural meanings, values, norms, expectations).[10] Shifting design focus toward these external variables helps us transition toward a more resource-efficient economy, supporting users in keeping products, components, and materials at their highest utility and value throughout their lifetimes. To design products that last, we must work toward a synthesis of two forms of durability: "physical" and "experiential." With the physical, products must objectively withstand use and possess the capacity to be maintained and repaired over time. With the experiential, products must subjectively engage users in deep and resonant ways to form meaningful associations. Of course, the two approaches do not exist in binary opposition, as they might first appear. For example, when a button comes loose on a shirt and eventually falls off (physical failure), if one does not care enough (experiential failure), the shirt will most likely never receive the simple repair it needs to remain in service.

A team of researchers in the Netherlands explores design-led approaches to product life extension and the circular economy. Together, Conny Bakker, Marcel den Hollander, Ed van Hinte, and Yvo Zijlstra describe five approaches to design that lasts, each comprising a carefully orchestrated blend of both experiential and physical factors: "classic long life" (e.g., high-quality products with a long useful life, like a Miele washing machine); "hybrid" (e.g., blending a long-life product with short-life consumables, like an inkjet printer with replaceable cartridges); "gap exploiter" (e.g., enabling the capture of leftover value from discarded products, like the RAM module from a blown CPU); "access" (e.g., designing for use rather than ownership, like Zipcar or Spotify); and "performance" (e.g., paying for what you use, as in a pay-as-you-go laundry service).[11] In each example, we see that the physical and the experiential are two elements of a codependent whole, working together like the individual threads of a braided rope. It is not about one or the other (experiential or physical) but rather about a synergy of the two. Through this synergy, we create new generations of experientially rich products that enhance our everyday lives while encouraging social material practices like maintenance, repair, upgrade, customization, and sharing.

Policy makers are becoming increasingly engaged in exploring the roles that government can play in steering businesses and consumers toward more sustainable forms of production and consumption. Over the past century, we have seen a slow increase in the introduction of environmental legislation. In the United States, Congress passed several key pieces of legislation (e.g., the Clean Air Act of 1963, the Clean Water Act of 1972, the Endangered Species Act of 1973, the Montreal Protocol of 1987). These, along with a host of other examples, represent the gradual awakening of governments—in affluent nations like the United States—to the rapidly scaling social and environmental issues facing us, issues that in time will bring societies and economies to their knees unless immediate action is taken. For the foreseeable future, this upward legislative trend is set to continue, but it must gather pace if it is to make sustainable design an imperative for business. Cofounder of design studio Dash Marshall, Bryan Boyer describes how government institutions aim to tackle complex, twenty-first-century problems with outmoded eighteenth-century principles. Today's challenge, he says, "lies in rewriting assumptions inherited from previous eras of prosperity and creating ways of working across silos and other arbitrary boundaries—and quickly."[12] As do most legislative tools, these remain fairly blunt instruments, often too clumsy for the task at hand. Like most regulatory measures, they operate on a penal system that fines, taxes, and imprisons those in breach, rather than a merit system that rewards, compensates, and recognizes those meeting the highest standards.

In the European Union, the dominant environmental policy focus relating to design has been on the introduction of readily quantifiable eco-efficiency measures such as rating energy performance or banning the use of certain hazardous materials. These are important things to regulate and monitor, but such an approach overlooks equally fundamental issues relating to the short-lived, throwaway nature of these "sustainable design" products. As Paul Dillinger, vice president of global product innovation at Levi Strauss & Co., argues: "Why on Earth would you try to use sustainability messaging as a mechanism to sell more product, when in fact the quantity of product being sold is precisely the problem?"[13] In the consumer electronics sector, the Waste Electrical and Electronic Equipment Directive (WEEE Directive) begins to tackle the issue of disposability. It was passed into EU law in 2013 and sets collection, recycling, and recovery targets for any discarded electronic product. The cost of this recovery falls on the producer, like a landfill tax, so that product disposal becomes a business expense, something to try to avoid. The design responses to this waste-reduction legislation have been limited. Responses have a tendency toward end-of-pipe strategies like ease of disassembly, better recycling processes, or product take-back initiatives, rather than any deeper strategies that look to extend the lifetimes of products themselves. Again, recycling and product take-back are important tactical moves for design, but in the end, they merely attend to the symptoms of the problem (waste), rather than addressing its underlying root causes (wastefulness). On a more positive note, one might argue that legislation such as the WEEE Directive has helped create the conditions for system-level transformations like the circular economy to flourish. In this way, things are slowly improving as policy makers begin to look beyond traditional eco-efficiency measures to steer commercial practices toward product and material systems that are both circular and long-lasting.

Since 2010 political commitment has been growing for promoting longer product lifetimes, with design for durability being one of the product policy issues stressed by the European Commission's Circular Economy Action Plan.[14] This growing political commitment is mirrored by a rapidly increasing commitment from industry—a shared focus on material and product longevity that will set the design agenda for many decades to come. In its report, the European Commission guarantees to "promote the reparability, upgradability, durability, and recyclability of products by developing product requirements relevant to the circular economy in its future work under the Ecodesign Directive."[15] In 2011, the government of the United Kingdom proposed an economy where resources are used sustainably through design for longer life, upgrading, reuse, or repair. It called for a seismic shift in thinking from the design and delivery of short-life products to that of longer-lasting products, user experiences, and services.[16] Similarly, and with a greater focus on the use phase, right-to-repair legislation has been introduced in more than half the states in the United States, giving consumers an alternative to costly brand-owned service departments when their products fail.[17] This vitally important policy move reopens the product repair space and hails the return of forgotten practices of maintenance and repair. More recently, in 2015, the French government passed legislation to penalize companies that intentionally use techniques to reduce the life spans of products to increase replacement rates.[18] This decisive piece of legislation made France the first country in the world to effectively ban "planned obsolescence" practices and the deliberate curtailment of product life spans by design. To help enforce this kind of law, the European Commission is developing plans for an independent testing program to detect planned obsolescence practices found in products and services.[19] Any producer caught doing this could be found guilty of a criminal offense, potentially facing imprisonment or fines of up to €300,000 depending on the severity of infringement.[20]

Despite these positive policy moves, their effect on the profession of design is questionable and has made limited difference to the practices and processes of design as a prospective, exploratory, and ultimately hopeful field. Design is not motivated by a fear of punishment, avoidance of fines, or operating subordinately within prescribed rules and regulations. Designers are not compliant, well-behaved people, motivated by the call to fall in line and fit in. Indeed, we are trained not to be, as we answer to a far greater calling: to propose, meddle, disrupt, and transform for the better. Designers are motivated by the opportunity to create meaningful stuff, not meaningless rubbish; to have a greater social and ecological purpose; to make things that matter to people and contribute positively to the world. Beyond obvious social and ecological benefits, design that lasts offers creative sustenance, enduring meaning, and a genuine purposeful practice to those who are willing to engage with it. For most of us, this is the reason we got into design in the first place: to make things that matter.

Ecologies of Interventions

Designers operate within a system that wants to change but does not know how. Most people I meet in design know that something is wrong and things need to change. In most cases, they lack a sufficiently well-articulated and compelling vision of a future state, and the creative tools and methods to work creatively toward it. Oftentimes the gap between the current state and the imagined future state seems too vast to span. This vastness leads designers to feel overwhelmed by the enormity of the challenge in moving through this to that. They tell me of burned-out colleagues and how they feel the same hopelessness. They speak of stressed-out managers who warily share their concerns but have become so entrenched in their own realities that they feel unable to make the scale of change they suspect is needed. Indeed, most of us "feel" the pressing need for change but struggle to imagine it, and the few who can imagine it often do not know how to plot an incremental path toward it.

Too often we hear of large-scale systems change, and incremental design moves, discussed as contrasting approaches. This is a mistake, as both scales of change are vital, and we cannot have one without the other. When understood as part of a choreographed pathway, these small-scale, iterative design moves make far more sense, like individual footsteps on a daylong hike: each one does not seem like much when viewed in isolation (a single stride), but when sequenced together as a string of interlinked moves (the overall hike), great distances are covered, and far-off destinations reached. And so seemingly small, incremental design moves can collectively pave the way toward preferred future states. Equally, without a clearly held vision of a preferred future state to work toward, small, iterative moves rarely amount to much. Often these cumulative design efforts lead to uncoordinated meanderings, through which significant energy, material, and time are wasted in taking us exactly nowhere. In this uncoordinated design approach, we are worse off than before we started, as we have covered no ground, only now we are fatigued, resource spent, and standing atop ruined, trampled earth. We confuse activity with accomplishment and believe that because we are "working hard," we must be achieving something. Yet this is rarely the case. Without a shared sense of direction and purpose, we find people pulling in opposite directions and achieving relatively little.

The environmental scientist, teacher, and writer Donella Meadows describes these coordinated series of design moves as "leverage points," optimal places to intervene in a system. Systems are unified conglomerations of interdependent parts, delineated by their spatial and temporal boundaries, influenced by their environments. Systems are dynamic, not static, and changing one part of a system usually affects other parts and the whole system. By applying the right kinds of leverage to specific points in a system, design can encourage change. Meadows describes how dynamic systems studies are usually not designed to predict what "will" happen. Rather, they are designed to explore what "might" happen, if several driving factors (design interventions) unfold in a range of different ways, at a range of different times.[21] As the systems scientist Peter Senge states, "Systems thinking is a context for seeing wholes. It is a framework for seeing interrelationships rather than things, for seeing patterns of change rather than static snapshots."[22]

We face an array of complex problems, from an aging population to climate change to intergenerational cycles of poverty. It can often seem that these challenges are insurmountable and that we lack the ability to make meaningful change.[23] Indeed, viewed in isolation, one-off design changes at the product level can seem inadequate. Yet when viewed together as an interconnected sequence of design moves, these multilevel approaches advance the power of design in driving systems-level change. Terry Irwin, director of the Transition Design Institute at Carnegie Mellon University, describes this as an "ecology of interventions."[24] Transition design is a form of research and practice that shapes and supports the design-led societal transition toward a sustainable future. Transition designers use living systems theory as an approach to understanding wicked problems and designing solutions to address them. Irwin explains how climate change, loss of biodiversity, depletion of natural resources, and the widening gap between rich and poor are just a few of the wicked problems that require designers to explore new, systems-level approaches to understanding and solving problems.[25] These kinds of complex problems, she says, must be addressed through ecologies of interventions, at multiple scales, which take place over short, medium, and long time horizons. The key is to understand that each design move (e.g., an urban bike-sharing station) is merely a step on a longer transition pathway toward a sustainable future (e.g., a carbon-neutral economy); the individual step is not the desired change but an important step in moving toward it.

This is the nature of systems change: it is incremental and manifests over time in both planned and unplanned ways. As Helsinki Design Lab, a strategic design and research group, cautions: "Modern society is now beginning to see—sometimes painfully—that the most critical challenges we face are also the ones that are most interconnected or systemic in nature. By expanding our understanding of systemic problems, we can better appreciate the principles that govern them, and the risks they pose to society."[26] Indeed, systems change is the only thing businesses and governments should be discussing at this point. As the sustainable fashion pioneer Timo Rissanen describes, "As long as overproduction is the most profitable path, brands are unlikely to hit the brakes and become truly sustainable."[27] He urges that huge system change, not just huge growth, is the only thing that will lead industry toward a more sustainable future.

Anna Birney, director of Forum for the Future's School of System Change, outlines five capabilities for designers to become effective change agents at the systems level. These relate to "diagnosis" (identifying optimal points to intervene in a system); "strategy" (devising iterative transition pathways); "innovation" (staging radical responses to systemic problems); "collaboration" (working with others to expand potential); and "leadership" (developing a shared vision of a preferred future).[28] The last point is critical, if we are to codevelop sufficiently radical, compelling, and inclusive future visions to design our way toward.

Earth's delicately balanced, interconnected systems are in constant flux, striving for stasis. Tides of living and nonliving matter ebb and flow throughout these systems, negotiating equilibrium and pushing toward balance. Leyla Acaroglu, founder of the Unschool of Disruptive Design, advocates the design of a world that works better for all. To do this, she argues, requires a systems-level approach that meets our needs, is regenerative to people and planet, and "takes and gives at the same time."[29] As she puts it: "Everything created must take from something else; this is the law of entropy. Every industrial action has a reaction, an impact, and we currently do not design these trade-offs into our products, systems, or services."[30] To become a pioneer of systems-level change, rather than a replicator of systems-level problems, Acaroglu argues for a deeper commitment to designers working—and thinking—within the carrying capacities of living systems. That is not to say that "artifice" is necessarily bad or something to be avoided. Rather, we must learn to see that our design actions are directly linked to Earth's living systems and always shall be. When working toward design-led systems change, these "links" form a central part of the process and fall firmly within the designer's scope of responsibility. Consequential connections between design decisions and living systems are not unfortunate side effects or unseen externalities. These "designed effects" are an equal part of the creative output; they contribute to the "hyperobject," as described earlier in the book. Daniel Christian Wahl describes how these forms of interconnectedness, unpredictability, and uncontrollability are key characteristics of all complex, dynamic systems. We need to focus less on predicting and controlling outcomes, and more on designing for positive emergence.[31] It is interesting to note that when the "emergent properties" are positive, we readily "own" them, describing these favorable outputs as very much a part of the design intent. In contrast, when we believe the consequences to be negative, we tend not to own them, and instead place responsibility on other actors in the system—policy makers, consumers, economics, or the weather, for example. To enable sustainable change, we must own *all* the outcomes—the favorable (positive externalities) and the unfavorable (negative externalities)—and become far more skilled in recognizing the difference between the two.

Experience Heavy, Material Light

Design that lasts takes us toward a society that gets by with fewer, better things. Indeed, simply having more stuff stopped making people happier years ago. What we need is an economy of "better," not "more." To enable this form of societal transition, designers must forge deeper understandings of why people discard products that still work, and how we can design products people keep for longer. These are complex sociotechnical challenges. Yet, through engaging with such relational and systemic problems, we place ourselves at the heart of the "trouble." Donna Haraway, a prominent feminist scholar in the field of science and technology studies, describes this situatedness as being "more conducive to the kind of thinking that provides the means to build more liveable futures."[32]

In a world where "no products" is not an option, we must instead consider a society that flourishes with fewer products. Not less meaning, less value, or less experience; just less stuff. This requires a shift to what I call an "experience heavy, material light" design sensibility, through which we seek to elicit the richest experience with the least material. Through this approach, we decouple material throughput with human fulfillment. This is a more appropriate design response than to push for total dematerialization—a world without things, just people in their "natural" habitats. This retrograde vision is wholly divorced from the contemporary reality we have, ourselves, striven to build these past centuries. We are here now; this is our new nature, and we must operate from this position in a purposeful, constructive, and future-oriented manner.

The industrial design luminary Dieter Rams advocates a puritanical, barely noticeable form of industrial design, in which the absence of detail creates space for meaning. This "less but better" approach has important parallels with the pursuit of a populace that possesses fewer but better things. In the late 1970s, Rams was becoming increasingly concerned by the state of the world around him, and with what he described as an impenetrable confusion of forms, colors, and noises; a world of excess things, excess detail, excess stimulation.[33] He described this overstimulation as "pollution" that overwhelmed and poisoned the mind, inhibiting our ability to fully experience the world and form meaningful associations with it. The British fashion revolutionary Vivienne Westwood speaks with great passion about the importance of design that lasts. She describes three interlinked steps: "Buy less, choose well, and make it last."[34] In 2015 her newly opened Paris store had these words splashed across the window, sending this message to her own customers and anyone passing by. Westwood is a fierce advocate of sustainable fashion and frequently laments the failures of capitalism and calls attention to the need to change our spending behaviors and address the ethical and environmental issues of the industry.[35] Her message is an important and balanced one, simultaneously acknowledging the need to own things alongside the need to change the practices and norms underpinning this ownership. Similarly, the curator and scholar Glenn Adamson expresses how "many children and adults are losing touch with the material objects that have nurtured human development for thousands of years. The objects are still here," he says, but "we seem to care less and know less about them."[36] He bemoans the problematic notion of a throwaway society, calling instead for a culture of fewer, better things, with which we have deeper, more complex relationships over longer periods of time.

The design profession is in dire need of a more compelling vision of its contemporary role and purpose: one that reinstates designers as agents of social and environmental transformation, not peddlers of world-breaking whimsy. This is not about "redesigning design," driving a "design redux," or some other grandiose, ivory tower reframing of what design ought to be. Rather, I simply propose an expansion of design, extending the capacity of designers to think on a more systemic level and to consider the life of designed things over far longer horizons of time. If you want to design things that last, do more of what helps, and less of what does not. That is to say: there are things you can do to increase the life span of a product, and there are things you can do to decrease it—as the pages of this book attest. So all design really needs to do is develop a better understanding of what helps, and what does not, and to redirect practices away from the unhelpful, toward the helpful. This involves a degree of letting go, but letting go is far easier when you have something appealing to grab onto with your newly freed hand. Adopting a "more like this, less like that" mind-set is helpful in iteratively steering thinking and practice toward design that lasts. It supports the gradual migration away from socially and environmentally harmful practices, toward more socially and environmentally helpful ones. It is not a binary state, in which you used to design short-life products, and now you design long-life products. Change generally does not happen like this. Rather, it is an ongoing process of change, characterized by continual course corrections and adjustments, as we gradually become more accomplished in delivering meaningful, longer-lasting designed experiences. This involves a building out of layers of understanding, which does not replace what was there before but adds to and advances it. The Hegelian notion of *aufheben* describes this transitional process as a paradoxical state of both preserving and changing, and eventually advancement, all at once. This is, after all, what it is to evolve and grow. As Ruha Benjamin, a Professor of African American Studies at Princeton University reminds us, one must "remember to imagine and craft the worlds you cannot live without, just as you dismantle the ones you cannot live within."[37]

Designers Who Last

It is dangerous and somewhat egotistical to claim that one is working to "save the planet." The planet, in terms of its physics, is a mighty and formidable place and certainly does not need saving by you; this is navel-gazing, anthropocentric nonsense. If the world needs anything, it is to be left the heck alone. Or, to put it more plainly: it needs saving from us, not by us. As the sustainability communications specialist Ed Gillespie argues, our "reckless and ignorant attitude relies on three all-too-comforting assumptions. That we know what is happening to our world. That we are in control. And that we have the right leadership to address our situation."[38] Gillespie goes on to expose a far less comfortable reality than the one we have imagined ourselves into, wherein we do not fully understand what is happening, are absolutely not in control, and our leadership fails us at almost every level.

In terms of sustainability, many designers feel a moral obligation to embrace it all and do it all. However, people do not actually work that way. At least, they do not work that way for long. If you try to embrace it all and do it all, you are setting yourself up to fail, as everything you do will fall short. Indeed, it cannot all be done; at least not by you alone. It is tough to maintain a high level of engagement with so many critical social and environmental issues all at once. We might refer to this as a "peanut butter" approach in which we have a certain amount of resource (e.g., energy, time, motivation) and we choose to spread it evenly, but thinly, over the available surface (e.g., a discipline, a set of issues). With this approach, everyone gets a bit, but not much at all. This is not what we need, a design discipline characterized by swaths of near-identical generalists who have only superficial understandings of all of it, but no deep expertise in any one area: fashion designers with no practical understanding of pattern, drape, and form; furniture designers devoid of any material sensibility, having never set foot in a workshop before. At its worst, design operates like this, an overstretched and underfocused domain in which creative potential is squandered through ambitious individuals trying to do too many things quite well, as opposed to doing just a few things brilliantly.

To thrive in design—and for designers to last—you need two parallel modes of thinking, which reflexively switch between the "general" and the "particular." In the general mode, you seek to develop a sense of all of it, to some extent. In doing so, you develop a good mental model of the theoretical terrain, its key features, what broadly characterizes each of them, and how they may or may not interlink. In the particular mode, you zoom in on just one of these features and drill down as deep as you can go. This "drill site" is where your expertise is primarily situated. That is not to say that you cannot establish several drill sites and put time and energy into each one. Indeed, most of us do have specialist knowledge, skills, and capabilities related to more than one area. However, you cannot drill all of it, at least not to any significant depth. These two parallel modes are codependent, with each mode requiring the other for an individual's design process to truly flourish. Critically reflective design practitioners continually engage in the reflexive to-and-fro between thinking in general and particular terms about their work, and the issues it connects with. Their lived experience of design is characterized by a continual questioning of the particular project context they are working in right now, and why that matters to the general contexts it is nested within—working on something highly specific while maintaining awareness of the systemic conditions in which that specificity resides.

Indeed, to acquire deep expertise, you must spend time in a place, be it an actual place like a city or a business or a figurative place like a knowledge domain or political issue. You must become intimate with it; ultimately, you must be a resident there, not a tourist hurriedly passing through. When we consider the troubling nature of our complex, wicked problems, it can be difficult to stay put and go deep. Well-intentioned designers frequently burn themselves out by trying to be all places at once. They try to deal with all of it but, in so doing, accomplish little or nothing beyond self-destruction. This is not how to create designers who last. In contrast, large businesses with an extensive population of highly specialized employees might do better at trying to consider "all of it," and working at depth through multiple points of intervention. But for an individual, it is unhealthy to think like this. You are not expected to do it all alone. The responsibilities of a global business differ dramatically from the responsibilities of an individual, and it is my guess that you read these words now as an individual. You might well be an individual who is part of a large organization, but in the end, you are an individual: an individual with skills, sensibilities, strengths, weaknesses, things you care about, and things you do not. This cognitive practice of zooming in and out of the general and the particular is a necessary condition for designers who are looking to enable systems-level change: an agile mind, capable of shifting scale from micro to macro, and thinking in a manner that considers the impacts of each design move, as the creative process unfolds. This agile, reflexive state of mind is a characteristic found in all designers working productively as agents of sustainable change.

Like a drop of ink landing on a clean white tissue, imagine yourself landing in this domain, somewhere, and slowly spreading to become a larger part of it. Have realistic expectations and do not expect to be good at all of this, or to be in a situation where you have full control over all of it. What matters is that you have a good map of the general terrain, and a specific part of it where you wish to reside. In short, you need to find your place and start there. You will know when you have found this place of belonging; it is where your skills, passions, and beliefs come alive, and where your unique worldview provides a fulcrum, against which you can leverage positive change in the world.

Acknowledgments

Writing this book has been an incredible journey, but I cannot take all the credit and certainly could not have done it alone. I am truly grateful for all the support that colleagues, friends, and family have given these past years.

Enormous thanks to the MIT Press for agreeing to publish this book. I am particularly grateful to Doug Sery and Noah Springer for their expert guidance throughout the process. Thanks also to series editors Ken Friedman and Erik Stolterman for including this book in the MIT Press Design Thinking, Design Theory collection; it is an honor to be part of it.

Much gratitude goes to my colleagues and students and Carnegie Mellon University's School of Design. It is a privilege to spend each day surrounded by such brilliant, caring people. Particular thanks go to TC Eley, Mary Tsai, Gautham Krishna, Corinne Britto, Christianne Francovich, and Matt Geiger. These graduate students provided phenomenal research assistance during the book's development, and for that I am grateful.

Finally, I owe so much of this to my wife, Ming Ming. She is the smartest person I know, and over the years has provided so much of the intellectual stimulus for this work. My son Jasper also deserves a nod for being such a wonderful kid and helping me see the world through another's eyes, eyes that are wholly good. With much appreciation and respect to you both—thank you.

Notes

Chapter 1

1. V. Papanek, *Design for the Real World* (London: Thames & Hudson, 1972), 3.

2. C. Haskins, "AirPods Are a Tragedy," *Vice*, May 6, 2019, https://www.vice.com/en_us/article/neaz3d/airpods-are-a-tragedy (accessed July 2, 2019).

3. Haskins, "AirPods Are a Tragedy."

4. Sherif, M., *The Psychology of Social Norms* (New York: Harper, 1936).

5. R. D. Cialdini, "Crafting Normative Messages to Protect the Environment," *Current Directions in Psychological Science* 12, no. 4 (2003): 105–109.

6. G. C. Bowker and S. L. Star, *Sorting Things Out: Classification and Its Consequences* (Cambridge, MA: MIT Press, 2000).

7. W. R. Stahel and T. Jackson, "Durability and Optimal Utilization: Product-Life Extension in the Service Economy," in *Clean Production Strategies*, ed. T. Jackson (Boca Raton, FL: Lewis, 1993), 261–294.

8. M. Stead, P. Coulton, and J. Lindley, "Spimes Not Things: A Design Manifesto for a Sustainable Internet of Things," EAD 2019, Proceedings, 2133–2152.

9. Federal Trade Commission, "Internet of Things: Privacy and Security in a Connected World," FTC Staff Report, https://www.ftc.gov/system/files/documents/reports/federal-trade-commission-staff-report-november-2013-workshop-entitled-internet-things-privacy/150127iotrpt.pdf (accessed August 9, 2019).

10. A. Escobar, *Designs for the Pluriverse: Radical Interdependence, Autonomy, and the Making of Worlds* (Durham, NC: Duke University Press, 2018), 32.

11. T. Fry, *New Design Philosophy: An Introduction to Defuturing* (Sydney: University of New South Wales Press, 1999).

12. E. Manzini and J. Cullars, "Prometheus of the Everyday: The Ecology of the Artificial and the Designer's Responsibility," *Design Issues* 9, no. 1 (Autumn 1992): 5–20.

13. Manzini and Cullars, "Prometheus of the Everyday."

14. H. Wieser, N. Tröger, and R. Hübner, "The Consumers' Desired and Expected Product Lifetimes," in *Product Lifetimes and the Environment Conference Proceedings, 17–19 June, 2015—Nottingham, UK*, ed. T. Cooper, N. Braithwaite, M. Moreno, and G. Salvia (Nottingham: Nottingham Trent University, CADBE, 2015).

15. M. Sarner, "The Age of Envy: How to Be Happy When Everyone Else's Life Looks Perfect," *Guardian*, October 9, 2018.

16. Sarner, "The Age of Envy."

17. R. Bocock, *Consumption* (Oxon: Routledge, 1993).

18. J. A. Dykstra, "Why Millennials Don't Want to Buy Stuff," Fast Company, July 13, 2012, https://www.fastcompany.com/1842581/why-millennials-dont-want-buy-stuff (accessed July 30, 2014).

19. B. Scott, *The Heretic's Guide to Global Finance: Hacking the Future of Money* (London: Pluto Press, 2013).

20. M. Shayler, "Deconstruction #2: Mobile Phone," in *The Great Recovery Project Report* (London: RSA, 2013), 7.

21. J. B. Schor, *The Overspent American: Why We Want What We Don't Need* (New York: Harper Perennial, 1999), 17.

22. P. T. Kilborn, "Splurge: Why Americans Can't Stop Buying Stuff," *New York Times*, June 21, 1998.

23. S. Rosenbloom, "But Will It Make You Happy?" *New York Times*, August 7, 2010.

24. S. Lyubomirsky, "Hedonic Adaptation to Positive and Negative Experiences," in *The Oxford Handbook of Stress, Health, and Coping*, ed. C. S. Carver (Oxford: Oxford University Press, 2010), 233–239.

25. F. Dostoyevsky, *The House of the Dead* (Mineola, NY: Dover Publications, 2004), 81.

26. Lyubomirsky, "Hedonic Adaptation to Positive and Negative Experiences."

27. G. Cupchick, "The Half-Life of a Sustainable Emotion: Searching for Meaning in Product Usage," in *The Routledge Handbook of Sustainable Product Design*, ed. J. Chapman (Oxon: Routledge, 2017), 25–40.

28. Cupchick, "Half-Life of a Sustainable Emotion."

29. K. M. Sheldon and S. Lyubomirsky, "The Challenge of Staying Happier: Testing the Hedonic Adaptation Prevention Model," *Personality and Social Psychology Bulletin* 36, no. 5 (February 23, 2012): 670–680.

30. T. D. Wilson and D. T. Gilbert, "Affective Forecasting: Knowing What to Want," *Current Directions in Psychological Science* 14, no. 3 (June 1, 2005): 131–134.

31. E. Suh, E. Diener, and F. Fujita, "Events and Subjective Well-Being: Only Recent Events Matter," *Journal of Personality and Social Psychology* 70, no. 5 (1996): 1091–1102.

32. E. C. Hirschman and M. B. Holbrook, "Hedonic Consumption: Emerging Concepts, Methods and Propositions," *Journal of Marketing* 46, no. 3 (Summer 1982): 92–101.

33. Lyubomirsky, "Hedonic Adaptation to Positive and Negative Experiences."

34. M. Wallendorf and E. J. Arnould, "My Favorite Things: A Cross-Cultural Inquiry into Object Attachment, Possessiveness, and Social Linkage," *Journal of Consumer Research* 14, no. 4 (March 1988): 531–547.

35. Y. S. Sherif and E. L. Rice, "The Search for Quality: The Case of Planned Obsolescence," *Microelectronics Reliability* 26, no. 1 (1986): 75–85.

36. W. Stahel, "Durability, Function and Performance," in *Longer Lasting Products: Alternatives to the Throwaway Society*, ed. T. Cooper (Farnham: Gower, 2010), 157–177.

37. N. Whiteley, "Toward a Throw-Away Culture: Consumerism, 'Style Obsolescence' and Cultural Theory in the 1950s and 1960s," *Oxford Art Journal* 10, no. 2 (1987): 3–27.

38. A. Twemlow, *Sifting the Trash: A History of Design Criticism* (Cambridge, MA: MIT Press, 2017), 25.

39. R. Loewy, "Personal Letter," *Times* (London), November 19, 1945.

40. Sherif and Rice, "The Search for Quality."

41. K. Y. White, "Time to Throw Away Our Throwaway Culture," *Eco Business*, February 24, 2016, 17.

42. Whiteley, "Toward a Throw-Away Culture."

43. W. C. Satyro et al., "Planned Obsolescence or Planned Resource Depletion? A Sustainable Approach," *Journal of Cleaner Production* 195 (September 2018): 744–752.

44. R. Smithers, "Fast Fashion: Britons to Buy 50m 'Throwaway Outfits' This Summer," *Guardian*, July 11, 2019.

45. T. Ziemer, "Why Consumer Products Are Designed to Fail," All About Circuits, August 31, 2015, https://www.allaboutcircuits.com/news/why-consumer-products-are-designed-to-fail (accessed April 4, 2018).

46. G. Vince, "The High Cost of Our Throwaway Culture," BBC Future, November 28, 2012, http://www.bbc.com/future/story/20121129-the-cost-of-our-throwaway-culture (accessed November 30, 2012).

47. Vince, "High Cost of Our Throwaway Culture."

48. D. Sayers, *Creed or Chaos? and Other Essays in Popular Theology* (London: Religious Book Club, 1948), 16.

49. M. O'Brien, "Consumers, Waste and the 'Throwaway Society' Thesis: Some Observations on the Evidence," *International Journal of Applied Sociology* 3, no. 2 (2013): 19–27.

50. O'Brien, "Consumers, Waste and the 'Throwaway Society.'"

51. D. J. C. MacKay, *Sustainable Energy without the Hot Air* (Cambridge: UIT Cambridge Ltd., 2009), 2.

52. J. Bruner, *Acts of Meaning: Four Lectures on Mind and Culture*, The Jerusalem-Harvard Lectures, book 3 (Cambridge, MA: Harvard University Press, 1990).

53. J. Aitchison, *The Language Web: The Power and Problem of Words* (Cambridge: Cambridge University Press, 1997), 130.

54. E. Oishi, "Semantic Meaning and Four Types of Speech Act," in *Perspectives on Dialogue in the New Millennium*, ed. P. Kühnlein, H. Rieser, and J. Benjamins (Amsterdam: John Benjamins, 2003).

55. R. W. Gibbs, "Understanding and Literal Meaning," *Cognitive Science* 13 (1989): 243–251.

56. D. Leontiev, *Positive Psychology in Search for Meaning* (Oxon: Routledge, 2014).

57. S. F. Pennock, "On the Meaning of Meaning: What Are We Really Looking For?" Positive Psychology, November 24, 2019, https://www.positivepsychologyprogram.com/meaning (accessed September 23, 2017).

58. P. Smagorinsky, "If Meaning Is Constructed, What's It Made From? Toward a Cultural Theory of Reading," *Review of Educational Research* 71, no. 1 (Spring 2001): 133–169.

Chapter 2

1. J. Dewey, *Theory of Valuation* (Chicago: University of Chicago Press, 1939).

2. M. Heidegger, *Being and Time* (New York: Harper Perennial Modern Classics, 2008).

3. M. Rokeach, *The Nature of Human Values* (New York: Free Press, 1973).

4. T. Adorno, *Minima Moralia: Reflections from Damaged Life*, Radical Thinkers (London: Verso, 2006).

5. D. J. Clandinin and F. M. Connelly, *Narrative Inquiry: Experience and Story in Qualitative Research* (San Francisco: Jossey-Bass, 2000).

6. S. Turkle, ed., *Evocative Objects: Things We Think With* (Cambridge, MA: MIT Press, 2007), 307.

7. A. Rose, "How to Build Something That Lasts 10,000 Years," BBC Future, June 11, 2019, https://www.bbc.com/future/story/20190611-how-to-build-something-that-lasts-10000-years (accessed June 11, 2019).

8. E. Chin, *My Life with Things: The Consumer Diaries* (Durham, NC: Duke University Press, 2016).

9. F. L. K. Hsu, "The Self in Cross-Cultural Perspective," in *Culture and Self: Asian and Western Perspectives*, ed. A. J. Marsella, G. deVos, and F. L. K. Hsu (New York: Tavistock Publications, 1985), 24–55.

10. M. Wallendorf and E. J. Arnould, "My Favorite Things: A Cross-Cultural Inquiry into Object Attachment, Possessiveness, and Social Linkage," *Journal of Consumer Research* 14, no. 4 (March 1988): 531–547.

11. F. Snare, "The Concept of Property," *American Philosophical Quarterly* 9, no. 2 (1972): 200–206.

12. J. L. Pierce, T. Kostova, and K. T. Dirks, "Toward a Theory of Psychological Ownership in Organizations," *Academy of Management Review* 26, no. 2 (2001): 298–310.

13. R. W. Belk, "Possessions and the Extended Self," *Journal of Consumer Research* 15, no. 2 (1988): 139–168.

14. W. Baxter and P. Childs, "Designing Circular Possessions," in *Routledge Handbook of Sustainable Product Design*, ed. J. Chapman (Oxon: Routledge, 2017), 391–404.

15. L. van Dyne and J. L. Pierce, "Psychological Ownership and Feelings of Possession: Three Field Studies Predicting Employee Attitudes and Organizational Citizenship Behaviour," *Journal of Organizational Behavior* 25, no. 4 (2004): 439–459.

16. G. McCracken, "Culture and Consumption: A Theoretical Account of the Structure and Movement of the Cultural Meaning of Consumer Goods," *Journal of Consumer Research* 13, no. 1 (1986): 71–84.

17. Baxter and Childs, "Designing Circular Possessions."

18. J. J. Arnett, "The Psychology of Globalization," *American Psychologist* 57, no. 10 (2002): 774–783.

19. C. Kluckhohn and W. H. Kelly, "The Concept of Culture," in *The Science of Man in the World Culture*, ed. R. Linton (New York, 1945), 78–105.

20. K. A. Zimmermann, "What Is Culture?" Live Science, July 13, 2017, https://www.livescience.com/21478-what-is-culture-definition-of-culture.html (accessed January 22, 2018).

21. P. Desmet and P. Hekkert, "Framework of Product Experience," *International Journal of Design* 1, no. 1 (2007): 57–66.

22. J. A. Banks and C. A. McGee, *Multicultural Education: Issues and Perspectives* (Needham Heights, MA: Allyn & Bacon, 1989).

23. Banks and McGee, *Multicultural Education*.

24. Wallendorf and Arnould, "My Favorite Things."

25. Desmet and Hekkert, "Framework of Product Experience."

26. Desmet and Hekkert, "Framework of Product Experience."

27. Desmet and Hekkert, "Framework of Product Experience."

28. R. Williams, "The Analysis of Culture," in *Cultural Theory and Popular Culture: A Reader*, ed. J. Storey (Athens: University of Georgia Press, 1998), 48–56.

29. W. Sieck, "Cultural Norms: Do They Matter?" Global Cognition, https://www.globalcognition.org/cultural-norms (accessed May 7, 2019).

30. D. Ravasi and V. Rindova, "Creating Symbolic Value: A Cultural Perspective on Production and Exchange," *SSRN* 111, no. 4 (May 2004).

31. P. Bourdieu, "The Forms of Capital," in *Handbook of Theory of Research for the Sociology of Education*, ed. J. G. Richardson (Westport, CT: Greenwood, 1986), 56.

32. V. P. Rindova, "Cultural Consumption and Value Creation in Consumer Goods Technology Industries," *Academy of Management Proceedings* 2007, no. 1 (2017): article 1.

33. Bourdieu, "The Forms of Capital."

34. Ravasi and Rindova, "Creating Symbolic Value."

35. Ravasi and Rindova, "Creating Symbolic Value."

36. Wallendorf and Arnould, "My Favorite Things."

37. Wallendorf and Arnould, "My Favorite Things."

38. M. Douglas and B. Isherwood, *The World of Goods: Towards an Anthropology of Consumption* (London: Allen Lane, 1979).

39. Wallendorf and Arnould, "My Favorite Things."

40. M. Csikszentmihalyi and E. Rochberg-Halton, *The Meaning of Things: Domestic Symbols and the Self* (Cambridge: Cambridge University Press, 1981), 239.

41. Douglas and Isherwood, *The World of Goods*.

42. A. D. Napier, *Making Things Better: A Workbook on Ritual, Cultural Values, and Environmental Behavior* (Oxford: Oxford University Press, 2013).

43. A. Escobar, *Designs for the Pluriverse: Radical Interdependence, Autonomy, and the Making of Worlds* (Durham, NC: Duke University Press, 2018).

44. Escobar, *Designs for the Pluriverse*.

45. Escobar, *Designs for the Pluriverse*, 204.

46. Restart Project, "First Protest for the Right to Repair in Brussels," Restart Project, https://www.therestartproject.org/restart-project/first-ever-protest (accessed December 2, 2018).

47. E. Matchar, "The Fight for the 'Right to Repair,'" *Smithsonian*, July 13, 2016, https://www.smithsonianmag.com/innovation/fight-right-repair-180959764/#f7D6EbUdOyRPt6VX.99 (accessed July 13, 2016).

48. Restart Project, "Restart Radio: Strong Links between Thrift and Innovation, Past and Present," https://therestartproject.org/podcast/thrift-innovation-past-present (accessed December 3, 2018).

49. M. Norton, D. Mochon, and D. Ariely, "The IKEA Effect: When Labor Leads to Love," *Journal of Consumer Psychology* 22, no. 3 (September 2011): 453–460.

50. Norton, Mochon, and Ariely, "The IKEA Effect."

51. L. Festinger, *A Theory of Cognitive Dissonance* (Stanford, CA: Stanford University Press, 1957).

52. Norton, Mochon, and Ariely, "The IKEA Effect."

53. N. Gregson, A. Metcalfe, and L. Crewe, "Practices of Object Maintenance and Repair: How Consumers Attend to Consumer Objects within the Home," *Journal of Consumer Culture* 9, no. 2 (June 2009): 248–272.

54. Gregson, Metcalfe, and Crewe, "Practices of Object Maintenance and Repair."

55. Gregson, Metcalfe, and Crewe, "Practices of Object Maintenance and Repair."

56. G. Keulemans, N. Rubenis, and A. Marks, "Object Therapy: Critical Design and Methodologies of Human Research in Transformative Repair," *PLATE Conference Proceedings*, Delft, November 8–10, 2017, 186–191.

57. My Modern Met, "Kintsugi: The Centuries-Old Art of Repairing Broken Pottery with Gold," *My Modern Met*, https://www.mymodernmet.com/kintsugi-kintsukuroi (accessed February 17, 2018).

58. My Modern Met, "Kintsugi."

59. E. Kalantidou, "Handled with Care: Repair and Share as Waste Management Strategies and Community Sustaining Practices," in *Product Lifetimes and the Environment Conference Proceedings, 17–19 June, 2015—Nottingham, UK*, ed. T. Cooper, N. Braithwaite, M. Moreno, and G. Salvia (Nottingham: Nottingham Trent University, CADBE, 2015), 158–165.

60. C. Thompson, "We Need a *Fixer* (Not Just a Maker) Movement," *Wired*, June 18, 2013, https://www.wired.com/2013/06/qq-thompson (accessed March 14, 2016).

61. L. Houston et al., "Values in Repair," *CHI '16: Proceedings of the 2016 CHI Conference on Human Factors in Computing Systems*, San Jose, CA, May 7–12, 2016, 1403–1414.

62. Thompson, "We Need a *Fixer*."

Chapter 3

1. R. Macfarlane, *Underland: A Deep Time Journey* (New York: W. W. Norton, 2019).

2. C. Haskins, "AirPods Are a Tragedy," *Vice*, May 6, 2019, https://www.vice.com/en_us/article/neaz3d/airpods-are-a-tragedy (accessed May 6, 2019), 29.

3. K. Crawford and V. Joler, *Anatomy of an AI System: The Amazon Echo as an Anatomical Map of Human Labor, Data and Planetary Resources* (New York: Share Lab/AI Now Institute, 2018), 6.

4. Crawford and Joler, *Anatomy of an AI System*, 7.

5. Crawford and Joler, *Anatomy of an AI System*, 11.

6. M. Ananny and K. Crawford, "Seeing without Knowing: Limitations of the Transparency Ideal and Its Application to Algorithmic Accountability," *New Media and Society* 20, no. 3 (2018): 973–989.

7. Crawford and Joler, *Anatomy of an AI System*, 4.

8. Crawford and Joler, *Anatomy of an AI System*, 6.

9. M. Berners-Lee, *How Bad Are Bananas? The Carbon Footprint of Everything* (London: Profile Books, 2010).

10. K. Davies and L. Young, *Tales from the Dark Side of the City: The Breastmilk of the Volcano Bolivia and the Atacama Desert Expedition* (London: Unknown Fields, 2016).

11. C. Fitzpatrick, "Conflict Minerals and the Politics of Stuff," in *The Routledge Handbook of Sustainable Product Design*, ed. J. Chapman (Oxon: Routledge, 2017), 197–205.

12. Fitzpatrick, "Conflict Minerals and the Politics of Stuff."

13. E. Pilkington, "Robin Wright Targets Congo's 'Conflict Minerals' Violence with New Campaign," *Guardian*, May 17, 2016.

14. P. Le Billion, "Getting It Done: Instruments of Enforcement," in *Natural Resources and Violent Conflict*, ed. I. Bannon and P. Collier (Washington, DC: World Bank, 2013), 215–286.

15. Fitzpatrick, "Conflict Minerals and the Politics of Stuff."

16. P. Gleick, "Water and Conflict," *International Security* 18, no. 1 (Summer 1993): 79–112.

17. M. Anderson, "DR Congo's Miners Bear Brunt of Attempts to Make Minerals Conflict-Free," *Guardian*, September 10, 2014.

18. S. Raghavan, "Obama's Conflict Minerals Law Has Destroyed Everything, Say Congo Miners," *Guardian*, December 2, 2014.

19. Pilkington, "Robin Wright Targets."

20. Pilkington, "Robin Wright Targets."

21. D. Aronson, "How Congress Devastated Congo," *New York Times*, August 7, 2011.

22. Fitzpatrick, "Conflict Minerals and the Politics of Stuff."

23. M. Medina, *The World's Scavengers: Salvaging for Sustainable Consumption and Production* (New York: Altamira Press, 2007).

24. J. A. Matthews and H. Tan, "Circular Economy: Lessons from China," *Nature: International Weekly Journal of Science*, March 23, 2016, https://www.nature.com/news/circular-economy-lessons-from-china-1.19593 (accessed March 28, 2018).

25. J. DiGangi and J. Strakova, "Toxic Toy or Toxic Waste: Recycling Pops into New Products," IPEN, October 2015, https://ipen.org/sites/default/files/documents/toxic_toy_or_toxic_waste_2015_10-en.pdf (accessed October 31, 2017).

26. J. Macleod, *Ends: Why We Overlook Endings for Humans, Products, Services and Digital; And Why We Shouldn't* (London: Joe Macleod, 2017).

27. A. G. Ferrufino, "Building a Foundation for More Responsible Gold Mining in Uganda," Fairphone, September 11, 2018, https://www.fairphone.com/en/2018/09/11/building-foundation-for-responsible-gold-mining-in-uganda (accessed September 18, 2018).

28. A. Rose, "How to Build Something That Lasts 10,000 Years," BBC Future, June 11, 2019, https://www.bbc.com/future/story/20190611-how-to-build-something-that-lasts-10000-years (accessed June 11, 2019).

29. M. Ballester, "From Ownership to Service: A New Fairphone Pilot Just for Companies," Fairphone, January 8, 2018, https://www.fairphone.com/en/2018/01/08/from-ownership-to-service-new-fairphone-pilot-for-companies (accessed January 8, 2018).

30. J. Ash, *Phase Media: Space, Time and the Politics of Smart Objects* (London: Bloomsbury, 2017).

31. J. W. Forrester, "Counterintuitive Behavior of Social Systems," *Theory and Decision* 2, no. 2 (January 1971): 109–140.

32. H. Dubberly and P. Pangaro, "Cybernetics and Design: Conversations for Action," DDO, November 1, 2015, https://www.dubberly.com/articles/cybernetics-and-design.html (accessed November 1, 2015).

33. A. Vergou, M. Wong, and J. Morgan, "It's Not about Milk," Goldsmiths College, University of London, https://www.themilkhasturned.com/about (accessed July 2, 2019).

34. J. B. Thompson, *Studies in the Theory of Ideology* (London: Polity, 1984), 6.

35. D. Lockton and S. Candy, "A Vocabulary for Visions in Designing for Transition," *Cuadernos, del Centro de Estudios de Diseño y Comunicación* 73 (2019): 27–49.

36. G. Bateson, *Mind and Nature: A Necessary Unity* (New York: E. P. Dutton, 1979), 65.

37. G. Bateson, *Steps to an Ecology of Mind: Collected Essays in Anthropology, Psychiatry, Evolution, and Epistemology* (Chicago: University of Chicago Press, 2000).

38. A. Korzybski, *Science and Sanity: A Non-Aristotelian System and Its Necessity for Rigour in Mathematics and Physics* (Oxford: APA, 1933), 749.

39. R. Mazé, "Politics of Designing Visions of the Future," *Journal of Future Studies* 23, no. 3 (March 2019): 23–38.

40. R. Wray, "In Just 25 Years, the Mobile Phone Has Transformed the Way We Communicate," *Guardian*, December 31, 2009.

41. R. Hayes, *Rethinking Society from the Ground Up* (London: Alliance for Sustainability and Prosperity, 2013).

42. A. Kukla, "Antirealist Explanations of the Success of Science," *Philosophy of Science* 63, no. 1 (1996): 298–305.

43. I. Stewart and J. Cohen, *Figments of Reality: The Evolution of the Curious Mind* (Cambridge: Cambridge University Press, 1997).

44. L. Baker, "How Fashion Can Stop Ruining the Planet," BBC, April 20, 2018, https://www.bbc.com/culture/story/20180419-how-fashion-can-stop-ruining-the-planet (accessed April 20, 2018).

45. E. Karana, "How Do Materials Obtain Their Meanings?" *Journal of the Faculty of Architecture Middle East Technical University* 27, no. 2 (2010): 271–285.

46. L. Wittgenstein, *Philosophical Investigations* (Englewood Cliffs, NJ: Prentice Hall, 1999), 131.

47. Karana, "How Do Materials Obtain Their Meanings?"

48. K. Krippendorff and R. Butter, "Semantics: Meanings and Contexts of Artifacts," in *Product Experience*, ed. H. Schifferstein and P. Hekkert (Amsterdam: Elsevier, 2008), 353–375.

49. P. Wright, J. Wallace, and J. McCarthy, "Aesthetics and Experience-Centered Design," *ACM Transactions on Computer-Human Interaction* 15, no. 4 (November 2008): article 18.

50. J. A. Boydston, *The Later Works of John Dewey*, vol. 1, *1925–1953: 1925, Experience and Nature* (Carbondale: Southern Illinois University Press, 2008).

51. E. Giaccardi and E. Karana, "Foundations of Materials Experience: An Approach for HCI," *Proceedings of the 33rd SIGCHI Conference on Human Factors in Computing Systems* (New York: ACM), 2447–2456.

52. P. Desmet, "Designing Emotions," PhD thesis, Faculty of Industrial Design Engineering, TU Delft, the Netherlands, 2002.

53. Karana, "How Do Materials Obtain Their Meanings?"

Chapter 4

1. A. Dunne and F. Raby, *Design Noir: The Secret Life of Electronic Objects* (London: Birkhauser, 2001), 5.

2. P. Desmet and P. Hekkert, "Framework of Product Experience," *International Journal of Design* 1, no. 1 (2007): 57–66.

3. Desmet and Hekkert, "Framework of Product Experience."

4. K. C. J. Overbeeke and S. A. G. Wensveen, "From Perception to Experience, from Affordances to Irresistibles," in *Proceedings of 2003 International Conference on Designing Pleasurable Products and Interfaces*, ed. B. Hannington and J. Forlizzi (Pittsburgh: ACM Press, 2003), 92–97.

5. Desmet and Hekkert, "Framework of Product Experience."

6. N. H. Frijda, *The Emotions: Studies in Emotion and Social Interaction* (Cambridge: Cambridge University Press, 1987).

7. K. R. Scherer, A. Shorr, and T. Johnstone, eds., *Appraisal Processes in Emotion: Theory, Methods, Research* (Oxford: Oxford University Press, 2001), 22.

8. Desmet and Hekkert, "Framework of Product Experience."

9. P. Hekkert, "Design Aesthetics: Principles of Pleasure in Product Design," *Psychology Science* 48, no. 2 (2006): 157–172.

10. M. Hassenzahl, "User Experience and Experience Design," in *The Encyclopedia of Human-Computer Interaction*, 2nd ed., Interaction Design Foundation, https://www.interaction-design.org/literature/book/the-encyclopedia-of-human-computer-interaction-2nd-ed (accessed July 10, 2019).

11. I. McGilchrist, *The Master and His Emissary: The Divided Brain and the Making of the Western World* (New Haven, CT: Yale University Press, 2009), 270.

12. K. Michaelian, *Mental Time Travel: Episodic Memory and Our Knowledge of the Personal Past* (Cambridge, MA: MIT Press, 2016), 35.

13. D. Rose, *Enchanted Objects: Design, Human Desire, and the Internet of Things* (New York: Scribner, 2014), 39.

14. Hassenzahl, "User Experience and Experience Design."

15. J. Heskett, *Toothpicks and Logos: Design in Everyday Life* (Oxford: Oxford University Press, 2003), 16.

16. T. Kasser et al., "Materialistic Values: Their Causes and Consequences," in *Psychology and Consumer Culture: The Struggle for a Good Life in a Materialistic World*, ed. T. Kasser and A. D. Kanner (Washington, DC: American Psychological Association, 2004), 11–28.

17. R. Belk, "Three Scales to Measure Constructs Related to Materialism: Reliability, Validity, and Relationships to Measures of Happiness," *Advances in Consumer Research* 11 (1984): 291–297.

18. M. L. Richins and S. Dawson, "A Consumer Values Orientation for Materialism and Its Measurement: Scale Development and Validation," *Journal of Consumer Research* 19 (1992): 308.

19. H. Dittmar and L. Pepper, "To Have Is to Be: Materialism and Person Perception in Working-Class and Middle-Class British Adolescents," *Journal of Economic Psychology* 15, no. 2 (1994): 233–251.

20. E. van Hinte, ed., *Eternally Yours: Visions on Product Endurance* (Rotterdam: 010 Publishers, 1997), 234.

21. L. Crewe, *The Geographies of Fashion: Consumption, Space, and Value* (London: Bloomsbury, 2017), 116.

22. B. Hood, *The Science of Superstition: How the Developing Brain Creates Supernatural Beliefs* (New York: HarperOne, 2010), 27.

23. R. A. Georges and M. O. Jones, *Folkloristics: An Introduction* (Bloomington: Indiana University Press, 1995), 122.

24. K. R. Foster and H. Kokko, "The Evolution of Superstitious and Superstition-like Behaviour," *Biological Sciences* 276, no. 1654 (2008): 31–37.

25. S. A. Vyse, *Believing in Magic: The Psychology of Superstition* (Oxford: Oxford University Press, 2000), 19–22,

26. Vyse, *Believing in Magic*.

27. R. Horton, "African Traditional Thought and Western Science: Part I. From Tradition to Science," *Africa: Journal of the International African Institute* 37, no. 1 (1967): 50–71.

28. A. Mishara, "Klaus Conrad (1905–1961): Delusional Mood, Psychosis and Beginning Schizophrenia," *Schizophrenia Bulletin* 36, no. 1 (2010): 9–13.

29. G. Lewis, "The Look of Magic," *Man: The Journal of the Royal Anthropological Institute* 21, no. 3 (September 1986): 414–437.

30. S. Vyse, "Do Superstitious Rituals Work?" *Skeptical Inquirer* 42, no. 2 (2018): 32–34.

31. P. Valdesolo, "Why 'Magical Thinking' Works for Some People," *Scientific American*, October 19, 2010, https://www.scientificamerican.com/article/superstitions-can-make-you (accessed January 6, 2018).

32. D. W. Winnicott, "Transitional Objects and Transitional Phenomena: A Study of the First Not-Me Possession," *International Journal of Psycho-analysis* 34, no. 2 (1953): 89–97.

33. S. Turkle, *Evocative Objects: Things We Think With* (Cambridge, MA: MIT Press, 2011).

34. D. Kelemen, "The Scope of Teleological Thinking in Preschool Children," *Cognition* 70 (1999): 241–272.

35. E. Arthurs, "Travelodge Research Shows Over a Third of British Adults Still Sleep with a Teddy Bear," CISION PR Newswire, https://www.prnewswire.com/news-releases/travelodge-research-shows-over-a-third-of-british-adults-still-sleep-with-a-teddy-bear-139427998.html (accessed May 13, 2016).

36. J. M. Grohol, "Do You Still Have a Security Blanket?" PsychCentral, https://www.psychcentral.com/blog/do-you-still-have-a-security-blanket (accessed October 20, 2018).

37. A. Cuthbertson, "I'm Not That Superstitious," Studying Religion in Culture, January 27, 2016, https://www.religion.ua.edu/blog/2016/01/27/im-not-that-superstitious (accessed March 12, 2017).

38. NASA, "Lucky Peanuts," NASA Solar System Exploration, https://solarsystem.nasa.gov/news/10022/lucky-peanuts (accessed June 28, 2019).

39. Cuthbertson, "I'm Not That Superstitious."

40. S. Cannon, *Popular Beliefs and Superstitions from Utah* (Salt Lake City: University of Utah Press, 1984).

41. E. E. Evans-Pritchard, *Witchcraft, Magic, and Oracles among the Azande* (Oxford: Clarendon Press, 1937).

42. M. Schippers and P. A. M. van Lange, "The Psychological Benefits of Superstitious Rituals in Top Sport," *SSRN*, ERIM Report Series Reference no. ERS-2005-071-ORG, December 2005.

43. J. St. George, "The Things They Carry: A Study of Transitional Object Use among U.S. Military Personnel during and after Deployment," Theses, Dissertations, and Projects, no. 973, Smith College, https://scholarworks.smith.edu/theses/973 (accessed January 3, 2018).

44. S. Li, "Skipping the 13th Floor," *Atlantic*, February 2015, https://www.theatlantic.com/technology/archive/2015/02/skipping-the-13th-floor/385448 (accessed March 22, 2016).

45. S. Wagner, "Haunted Possessions," Live About, https://www.liveabout.com/haunted-possessions-2596712 (accessed July 7, 2019).

46. B. Hood, *The Science of Superstition: How the Developing Brain Creates Supernatural Beliefs* (New York: HarperOne, 2010), ix.

47. W. Baxter and P. Childs, "Designing Circular Possessions," in *Routledge Handbook of Sustainable Product Design*, ed. J. Chapman (Oxon: Routledge, 2017), 391–404.

48. S. Fokkinga, "Design –|+ Negative Emotions for Positive Experiences," PhD diss., TU Delft, the Netherlands, 2015.

49. M. Hassenzahl, *Experience Design: Technology for All the Right Reasons*, Synthesis Lectures on Human-Centered Informatics (San Francisco: Morgan and Claypool, 2010).

50. S. Fokkinga and P. Desmet, "Darker Shades of Joy: The Role of Negative Emotion in Rich Product Experiences," *Design Issues* 28, no. 4 (October 2012): 42–56.

51. M. Tsai, "Making Mistakes," graduate thesis, Carnegie Mellon University, School of Design, Pittsburgh, 2019.

52. B. Ehrenreich, *Bright-Sided: How Positive Thinking Is Undermining America* (London: Picador, 2010).

53. Fokkinga and Desmet, "Darker Shades of Joy."

54. P. F. Grendler, "The Enigma of 'Wisdom' in Pierre Charron," *Romance Notes* 4, no. 1 (Autumn 1962): 46–50.

55. Fokkinga, "Design –|+ Negative Emotions for Positive Experiences."

56. T. Ingold, "Towards an Ecology of Materials," *Annual Review of Anthropology* 41, no. 1 (2012): 427–442.

57. M. Ashby and K. Johnson, *Materials and Design: The Art and Science of Material Selection in Product Design* (Oxford: Butterworth-Heinemann, 2002), 73.

58. S. Fokkinga, P. Desmet, and J. Hoonhout, "The Dark Side of Enjoyment: Using Negative Emotions to Design for Rich User Experiences," *Proceedings—Design and Emotion 2010*, Chicago, 2010, https://www.zenodo.org/record/2596843#.XUrMA JNKjdd (accessed May 5, 2019).

59. Fokkinga, Desmet, and Hoonhout, "The Dark Side of Enjoyment."

60. Fokkinga, "Design –|+ Negative Emotions for Positive Experiences."

61. Michael J. Apter, *The Experience of Motivation: The Theory of Psychological Reversals* (New York: Academic Press, 1982).

62. A. Dunne and F. Raby, *Speculative Everything: Design, Fiction, and Social Dreaming* (Cambridge, MA: MIT Press, 2013).

Chapter 5

1. H. Arendt, *The Human Condition* (Chicago: University of Chicago Press, 1998), 138.

2. J. Wood, "The Culture of Academic Rigour: Does Design Research Really Need It?" *Design Journal* 3, no. 1 (2000): 44–57.

3. Arendt, *The Human Condition*, 137.

4. D. Lilley et al., "Cosmetic Obsolescence? User Perceptions of New and Artificially Aged Materials," *Materials and Design* 101 (April 2016): 355–365.

5. Lilley et al., "Cosmetic Obsolescence?"

6. Lilley et al., "Cosmetic Obsolescence?"

7. O. Pedgley et al., "Embracing Material Surface Imperfections in Product Design," *International Journal of Design* 12, no. 3 (2018): 21–23.

8. M. Zhang, "This Leica M-P 'Correspondent' Edition Was Designed by Lenny Kravitz," PetaPixel, February 24, 2015, https://www.petapixel.com/2015/02/24/this-leica-m-p-correspondent-edition-was-designed-by-lenny-kravitz (accessed August 11, 2016).

9. W. Odom and J. Pierce, "Improving with Age: Designing Enduring Interactive Products," *CHI '09 Extended Abstracts*, CHI '09 Conference on Human Factors in Computing Systems, Boston, April 4–9, 2009, 3793–3798.

10. E. Giaccardi et al., "Growing Traces on Objects of Daily Use: A Product Design Perspective for HCI," *DIS '14 Proceedings of the 2014 Conference on Designing Interactive Systems*, Vancouver, June 21–25, 2014, 473–482.

11. E. Karana, E. Giaccardi, and V. Rognoli, "Materially Yours," in *Routledge Handbook of Sustainable Product Design*, ed. J. Chapman (Oxon: Routledge, 2017), 206–221.

12. L. Koren, *Wabi-Sabi for Artists, Designers, Poets and Philosophers* (Point Reyes, CA: Imperfect Publishing, 2008), 7.

13. M. Oster, *Reflections of the Spirit: Japanese Gardens in America* (London: Dutton Studio Books, 1994).

14. B. Leach and S. Yanagi, *The Unknown Craftsman: A Japanese Insight into Beauty* (Tokyo: Kodansha International, 2013).

15. Koren, *Wabi-Sabi for Artists*.

16. A. Juniper, *Wabi-Sabi: The Japanese Art of Impermanence* (North Clarendon, VT: Tuttle Publishing, 2003).

17. R. Griggs Lawrence, "Wabi-Sabi: The Art of Imperfection," *Utne Reader*, September–October 2001, https://www.utne.com/mind-and-body/wabi-sabi (accessed December 29, 2016).

18. R. R. Powell, *Wabi Sabi Simple: Create Beauty. Value Imperfection. Live Deeply* (Adams Media, 2004).

19. S. Reid-Henry, "Arturo Escobar: A Post-development Thinker to Be Reckoned With," *Guardian*, November 5, 2012.

20. T. Ingold, "The Temporality of the Landscape," *World Archaeology* 25, no. 2 (1993): 152–174.

21. R. Small, "Being, Becoming, and Time in Nietzsche," in *The Oxford Handbook of Nietzsche*, ed. J. Richardson and K. Gemes (Oxford: Oxford University Press, 2013), 122–134.

22. M. Merleau-Ponty, *Phenomenology of Perception* (Oxon: Routledge, 2005), 363.

23. Merleau-Ponty, *Phenomenology of Perception*, 159.

24. C. Darwin, *On the Origin of Species: By Means of Natural Selection* (London: Dover Publications, 2006).

25. A. Dunne and F. Raby, *Speculative Everything: Design, Fiction, and Social Dreaming* (Cambridge, MA: MIT Press, 2013), 69.

26. F. E. Peters, *Greek Philosophical Terms: A Historical Lexicon* (New York: NYU Press, 1967), 178.

27. J. Chapman and B. Marmont, "The Temporal Fallacy: Design and Emotional Obsolescence," in *Routledge Handbook of Sustainable Design*, ed. R. B. Egenhoefer (Oxon: Routledge, 2018).

28. G. Harman, *Tool-Being: Heidegger and the Metaphysics of Objects* (Peru, IL: Open Court, 2002), 2.

29. T. Morton, *Hyperobjects: Philosophy and Ecology after the End of the World* (Minneapolis: University of Minnesota Press, 2013), 2.

30. E. Saarinen, *The Search for Form in Art and Architecture* (New York: Dover, 1985), 72.

31. J. Ash, *Phase Media: Space, Time and the Politics of Smart Objects* (London: Bloomsbury, 2017).

32. J. McCarthy and P. Wright, "Technology as Experience," *Interactions* 11, no. 5 (October 2004): 42–43.

33. McCarthy and Wright, "Technology as Experience."

34. McCarthy and Wright, "Technology as Experience."

35. D. Hill, *Dark Matter and Trojan Horses: A Strategic Design Vocabulary* (Helsinki: Strelka Press, 2014), 12.

36. G. T. Miller, *Environmental Science: Sustaining the Earth* (London: Wadsworth Publishing, 1993), 44.

37. S. Bayley, "What Makes a Design Classic?" *Independent*, August 27, 1999.

38. G. P. Lathrop, introduction to *The Scarlet Letter*, by N. Hawthorne (Scotts Valley, CA: CreateSpace Independent Publishing Platform, 2018), iv.

39. C. Tonkinwise, "Beauty in Use," *Design Philosophy Papers* 1, no. 2 (2003): 73–82.

40. Chapman and Marmont, "The Temporal Fallacy."

41. G. Agamben, *State of Exception* (Chicago: University of Chicago Press, 2005), 23.

42. T. Vardouli, "Making Use: Attitudes to Human-Artifact Engagements," *Design Studies* 41, part A (November 2015): 137–161.

43. Bayley, "What Makes a Design Classic?"

44. V. Rognoli and E. Karana, "Towards a New Materials Aesthetic Based on Imperfection and Graceful Ageing," in *Materials Experience: Fundamentals of Materials and Design*, ed. E. Karana, O. Pedgley, and V. Rognoli (Oxford: Butterworth-Heinemann, 2014), 145–154.

45. N. Foster, "The Future of Things," Hello Fosta https://www.hellofosta.com/writing/the-future-of-things (accessed July 28, 2019).

46. D. K. Rosner et al., "Designing with Traces," *CHI '13 Proceedings of the SIGCHI Conference on Human Factors in Computing Systems*, Paris, April 27–May 2, 2013, 1649–1658.

47. D. Diderot, "Regrets for My Old Dressing Gown, or A Warning to Those Who Have More Taste than Fortune," trans. M. Abidor for marxists.org (2005), in *Oeuvres complètes*, vol. 4 (Paris: Garnier Frères, 1875).

48. J. B. Schor, *The Overspent American: Why We Want What We Don't Need* (New York: Harper Perennial, 1999).

Chapter 6

1. C. B. Macpherson, *The Political Theory of Possessive Individualism: From Hobbes to Locke*, reprint ed. (Oxford: Oxford University Press, 1962), 210.

2. D. Stephens, "50 Things Every Man Should Own to Win at Life," TAM, https://www.theadultman.com/live-and-learn/things-every-man-should-own (accessed July 16, 2019).

3. P. Bourdieu, *Distinction: A Social Critique of the Judgement of Taste* (Oxford: Routledge, 1984).

4. C. Palahniuk, *Fight Club* (New York: W. W. Norton, 2005), 23.

5. M. Macvean, "For Many People, Gathering Possessions Is Just the Stuff of Life," *Los Angeles Times*, March 21, 2014.

6. Macvean, "For Many People."

7. Waste Resource Action Programme (WRAP), "WRAP Reveals the UK's £30 Billion Unused Wardrobe," WRAP, https://www.wrap.org.uk/content/wrap-reveals-uks-%C2%A330-billion-unused-wardrobe (accessed August 7, 2014).

8. R. Botsman, *What's Mine Is Yours: The Rise of Collaborative Consumption* (New York: Collins, 2011), 30.

9. J. Chapman, *Emotionally Durable Design: Objects, Experiences and Empathy* (Oxon: Routledge, 2015).

10. J. Owen, "'Possession Purgatory': Situated Divestment Delay in the Lifecycle of Objects," paper presented at "Making and Mobilising Objects: People, Process and Place," University of Warwick, Interdisciplinary Postgraduate Conference, University of Warwick, Leicester, February 21, 2015.

11. J. Owen, "Waiting to Reach This Lull When the Sadness Has Become Slightly Less Desperate: Dealing with Material A/Effects: Distancing Memories and Emotion in Self-Storage," paper presented at "The Senses and Spaces of Death, Dying and Remembering: Historical and Contemporary Perspectives," conference, Leeds, March 27–28, 2018.

12. Owen, "Possession Purgatory."

13. V. Gill, "Millions of Old Gadgets 'Stockpiled in Drawers,'" *Guardian*, August 21, 2019.

14. G. Kossoff, "Cosmopolitan Localism: The Planetary Networking of Everyday Life in Place," *Cuaderno 73, Transition Design Monograph* 19 (July 2018): 51–66.

15. H. W. J. Rittel and M. M. Webber, "Dilemmas in a General Theory of Planning," *Policy Sciences* 4, no. 2 (1973): 155–169.

16. Kosoff, "Cosmopolitan Localism."

17. Kosoff, "Cosmopolitan Localism."

18. A. L. Tsing, *The Mushroom at the End of the World: On the Possibility of Life in Capitalist Ruins* (Princeton, NJ: Princeton University Press, 2015).

19. D. Baker-Brown, *The Re-use Atlas: A Designer's Guide towards a Circular Economy* (London: RIBA Publishing, 2017), 13.

20. L. Kok, G. Wurpel, and A. Ten Wolde, *Unleashing the Power of the Circular Economy* (Amsterdam: IMSA, 2013).

21. Kok, Wurpel, and Ten Wolde, *Unleashing the Power*.

22. J. Boehnert, *Design, Ecology, Politics: Towards the Ecocene* (London: Bloomsbury, 2018), 7.

23. T. Morton, "The Mesh," in *Environmental Criticism for the Twenty-First Century*, ed. S. LeMenager, T. Shewry, and K. Hiltner (Oxon: Routledge, 2012), 19–30.

24. T. Ingold, *Making* (Oxon: Routledge, 2013), 132.

25. Ingold, *Making*, 133.

26. M. Guber, M. McDonough, M. Kausch, S. Glass, and A. Gullingsrud, "Safe and Circular by Design: Making Positive Material Choices," ThinkDIF, https://www.thinkdif.co/sessions/safe-circular-by-design-making-postive-material-choices (accessed January 7, 2019).

27. SPACE10, "Closing the Loop: Welcome to the Circular Economy," chap. 6 of *Imagine*, Medium, July 25, 2017, https://www.medium.com/space10-imagine/chapter-6-closing-the-loop-welcome-to-the-circular-economy-92665c9678d9 (accessed August 20, 2017).

28. M. Perella, "Durable Duds: The Disruptive Startups Looking to Wear Out Fast Fashion," Sustainable Brands, https://www.sustainablebrands.com/read/product-service-design-innovation/durable-duds-the-disruptive-startups-looking-to-wear-out-fast-fashion (accessed May 2, 2018).

29. RSA: The Great Recovery, "Designing for a Circular Economy," RSA, https://www.greatrecovery.org.uk/resources/designing-for-a-circular-economy (accessed May 23, 2016).

30. M. Merleau-Ponty, *Phenomenology of Perception* (Oxon: Routledge, 1962), 222.

31. S. Misra and J. Maxwell, "Three Keys to Unlocking Systems-Level Change," *Stanford Social Innovation Review*, April 29, 2016, https://www.ssir.org/articles/entry/three_keys_to_unlocking_systems_level_change (accessed June 3, 2019).

32. H. Dubberly, U. Haque, and P. Pangaro, "What Is Interaction? Are There Different Types?" Dubberly Design Office, http://www.dubberly.com/articles/what-is-interaction.html (accessed January 7, 2019).

33. Dubberly, Haque, and Pangaro, "What Is Interaction?"

34. M. Moritz, "Open Property Regimes," *International Journal of the Commons* 10, no. 2 (2016): 688–708.

35. S. Sassen, "Who Owns Our Cities—and Why This Urban Takeover Should Concern Us All," *Guardian*, November 24, 2015.

36. W. F. Lloyd, *Two Lectures on the Checks to Population* (Oxford: Oxford University, 1833).

37. G. Hardin, "The Tragedy of the Commons," *Science* 162, no. 3859 (1968): 1243–1248.

38. K. Crawford and V. Joler, *Anatomy of an AI System: The Amazon Echo as an Anatomical Map of Human Labor, Data and Planetary Resources* (New York: Share Lab/AI Now Institute, 2018), 13.

39. J. Gapper, "Facebook Faces the Tragedy of the Commons," *Financial Times*, November 29, 2017.

40. Gapper, "Facebook Faces the Tragedy of the Commons."

41. Macpherson, *Political Theory of Possessive Individualism*.

42. C. Jarrett, "Why Are We So Attached to Our Things?" TED Talk, https://www.scienceandnonduality.com/why-are-we-so-attached-to-our-things (accessed January 30, 2017).

43. A. B. Weiner, *Inalienable Possessions* (Berkeley: University of California Press, 1992).

44. R. Belk, "Why Not Share Rather than Own?" *Annals of the American Academy of Political and Social Science* 611, no. 1 (May 2007): 126–140.

45. L. Downe, "Good Services Are Verbs, Bad Services Are Nouns," Design in Government, June 22, 2015, https://www.designnotes.blog.gov.uk/2015/06/22/good-services-are-verbs-2 (accessed July 4, 2017).

46. D. Sacks, "The Sharing Economy," Fast Company, https://www.fastcompany.com/1747551/sharing-economy (accessed October 7, 2013).

47. A. Rinne, "The Dark Side of the Sharing Economy," World Economic Forum, January 2018, https://www.weforum.org/agenda/2018/01/the-dark-side-of-the-sharing-economy (accessed September 15, 2018).

48. Rinne, "Dark Side of the Sharing Economy."

49. S. Troncoso, "Is Sharewashing the New Greenwashing?" *P2P Foundation Blog*, https://blog.p2pfoundation.net/is-sharewashing-the-new-greenwashing/2014/05/23 (Accessed July 16, 2017).

50. A. Light and C. Miskelly, "Sharing Economy vs. Sharing Cultures? Designing for Social, Economic and Environmental Good," *Interaction Design and Architecture(s)* 24 (2015): 49–62.

51. Light and Miskelly, "Sharing Economy vs. Sharing Cultures?"

52. A. Dall and G. Smith, "The Sympathy of Things," BBC Sounds, https://www.bbc.co.uk/sounds/play/m0001188 (accessed April 21, 2019).

53. Dall and Smith, "The Sympathy of Things."

54. Dall and Smith, "The Sympathy of Things."

55. B. J. Pine, *Mass Customization: The New Frontier in Business Competition* (Boston: Harvard Business Review Press, 1992), 307.

56. SPACE10, "Putting the 'Fab' in Fabrication: Manufacturing in the Digital Age," chap. 1 of *Imagine*, Medium, June 15, 2017, https://www.medium.com/space10-imagine/chapter-1-putting-the-fab-in-fabrication-manufacturing-in-the-digital-age-fc9c7670dc5c (accessed August 20, 2017).

57. Unmade, "Our Vision," Unmade, https://www.unmade.com/vision/# (accessed August 5, 2019).

58. D. Gauntlett, *Making Is Connecting* (Cambridge: Polity, 2011).

59. R. Phillips, *Futurekind: Design by and for the People* (London: Thames & Hudson, 2019).

60. Phillips, *Futurekind*, 17.

61. M. Ames, *The Charisma Machine: The Life, Death, and Legacy of One Laptop per Child* (Cambridge, MA: MIT Press, 2019).

62. T. Thwaites, *The Toaster Project, or A Heroic Attempt to Build a Simple Electric Appliance from Scratch* (Hudson, NY: Princeton Architectural Press, 2011), 182.

63. Restart Project, "Move Slow and Fix Things," https://www.therestartproject.org/about (accessed February 26, 2017).

Chapter 7

1. E. Ellis, "Anthropocene," The Encyclopedia of Earth, https://editors.eol.org/eoearth/wiki/Anthropocene (accessed May 5, 2019).

2. J. Macleod, *Ends: Why We Overlook Endings for Humans, Products, Services and Digital; And Why We Shouldn't* (London: Joe Macleod, 2017).

3. T. Cooper, "Which Way to Turn? Product Longevity and Business Dilemmas in the Circular Economy," in *Routledge Handbook of Sustainable Product Design*, ed. J. Chapman (Oxon: Routledge, 2017), 405–422.

4. J. Maynard Keynes, *The General Theory of Employment, Interest, and Money* (Seattle: Stellar Classics, 2016), 17.

5. Cooper, "Which Way to Turn?"

6. Nike, "Circularity: Guiding the Future of Design," https://www.nikecirculardesign.com (accessed August 1, 2019).

7. E. Thomasson, "IKEA to Test Furniture Rental in 30 Countries," Reuters Sustainable Business, https://www.reuters.com/article/us-ikea-sustainability-idUSKCN1RF0WY (accessed April 7, 2019).

8. Waste and Resource Action Programme (WRAP), *Valuing Our Clothes: The Evidence Base*, technical report (Banbury: WRAP, 2012).

9. Global Fashion Agenda, "Pulse of the Fashion Industry 2019," GFA, https://globalfashionagenda.com/initiatives/pulse/# (accessed May 29, 2019).

10. Cooper, "Which Way to Turn?"

11. C. Bakker et al., *Products That Last: Product Design for Circular Business Models* (TU Delft Library, 2014).

12. B. Boyer et al., *In Studio: Recipes for Systemic Change* (Helsinki: Sitra, 2011), 22.

13. J. Zientek, "Can Better Denim Change the World? Levi's Is Betting on It," Gear Patrol, https://www.gearpatrol.com/2019/08/08/paul-dillinger-levis-innovation (accessed August 10, 2019).

14. C. Dalhammar and J. Luth Richter, "Options for Lifetime Labeling: Design, Scope and Consumer Interfaces," in *Product Lifetimes and the Environment Conference Proceedings, 17–19 June, 2015—Nottingham, UK*, ed. T. Cooper, N. Braithwaite, M. Moreno, and G. Salvia (Nottingham: Nottingham Trent University, CADBE, 2015), 461–463.

15. European Commission, "Closing the Loop: An EU Action Plan for the Circular Economy," *Communication from the Commission to the European Parliament, the Council, the European Economic and Social Committee and the Committee of the Regions* (Brussels, 2015), 4

16. Her Majesty's Government, *Our Waste, Our Resources: A Strategy for England* (London: Defra, Resources and Waste Strategy Team, 2018).

17. M. Sullivan, "'Right to Repair' Legislation Has Now Been Introduced in 17 States," Fast Company, https://www.fastcompany.com/40518779/right-to-repair-legislation-has-now-been-introduced-in-17-states (accessed February 14, 2018).

18. J. Valant, "Planned Obsolescence: Exploring the Issue," European Parliamentary Research Service, 2016, https://www.europarl.europa.eu/RegData/etudes/BRIE/2016/581999/EPRS_BRI(2016)581999_EN.pdf (accessed July 27, 2017).

19. Cooper, "Which Way to Turn?"

20. A. Michel, "Product Lifetimes through the Various Legal Approaches within the EU Context: Recent Initiatives against Planned Obsolescence," *Proceedings*, PLATE Conference, Delft University of Technology, Delft, November 8–10, 2017, 266–269.

21. D. H. Meadows, *Thinking in Systems: A Primer* (London: Earthscan, 2010).

22. P. M. Senge, *The Fifth Discipline: The Art and Practice of a Learning Organisation* (London: Random House, 1990), 7.

23. R. Conway, I. Masters, and J. Thorold, *From Design Thinking to Systems Change: How to Invest in Innovation for Social Impact* (London: RSA Action and Research Centre, 2007).

24. T. Irwin, "The Emerging Transition Design Approach," *Proceedings*, Design Research Society Conference 2018, Limerick, June 25–28, 2018, 968–989.

25. T. Irwin, "Transition Design Preface," *Cuaderno 73*, *Transition Design Monograph* 19 (July 2018): 19–26.

26. Boyer et al., *In Studio*, 19.

27. S. Benson, "The Myth of Sustainable Fashion," *Huffington Post*, February 7, 2020.

28. A. Birney, "System Change Capabilities: What Are the Capabilities We Need for System Change?" School of System Change, Forum for the Future, https://www.forumforthefuture.org/system-change-capabilities (accessed November 23, 2018).

29. L. Acaroglu, "A Manifesto for Design-Led Systems Change," Disruptive Design, https://www.medium.com/disruptive-design/a-manifesto-for-design-led-systems-change-28ac240db6dd (accessed February 6, 2018).

30. Acaroglu, "Manifesto."

31. D. C. Wahl, "A Brief History of Systems Science, Chaos and Complexity," Age of Awareness, https://www.medium.com/age-of-awareness/a-brief-history-of-systems-science-chaos-and-complexity-d9198b1a198d (accessed July 10, 2019).

32. D. Haraway, *Staying with the Trouble: Making Kin in the Chthulucene* (Durham, NC: Duke University Press, 2016), 40.

33. D. Rams, *Less but Better* (New York: Gestalten, 2014).

34. K. Grant, "Vivienne Westwood: Everyone Buys Too Many Clothes," *Daily Telegraph*, September 16, 2013.

35. M. Hill, "How to Buy Less, Choose Well and Make It Last," Good on You, https://www.goodonyou.eco/how-to-buy-less-choose-well-and-make-it-last (accessed August 1, 2019).

36. G. Adamson, *Fewer, Better Things: The Hidden Wisdom of Objects* (London: Bloomsbury, 2018), 5.

37. Benjamin, R., *Captivating Technology: Race, Carceral Technoscience, and Liberatory Imagination in Everyday Life* (Durham, NC: Duke University Press, 2019), 11.

38. E. Gillespie, "The End of 'Saving the World'?" Medium, https://medium.com/@edgillespie2018/the-end-of-saving-the-world-3f7c00d5338c (accessed February 14, 2020).

Bibliography

Acaroglu, L. "A Manifesto for Design-Led Systems Change." Disruptive Design. https://www.medium.com/disruptive-design/a-manifesto-for-design-led-systems-change-28ac240db6dd (accessed February 6, 2018).

Adamson, G. *Fewer, Better Things: The Hidden Wisdom of Objects*. London: Bloomsbury, 2018.

Adorno, T. *Minima Moralia: Reflections from Damaged Life*. Radical Thinkers. London: Verso, 2006.

Agamben, G. *State of Exception*. Chicago: University of Chicago Press, 2005.

Aitchison, J. *The Language Web: The Power and Problem of Words*. Cambridge: Cambridge University Press, 1997.

Ames, M. *The Charisma Machine: The Life, Death, and Legacy of One Laptop per Child*. Cambridge, MA: MIT Press, 2019.

Ananny, M., and K. Crawford. "Seeing without Knowing: Limitations of the Transparency Ideal and Its Application to Algorithmic Accountability." *New Media and Society* 20, no. 3 (2018): 973–989.

Anderson, M. "DR Congo's Miners Bear Brunt of Attempts to Make Minerals Conflict-Free." *Guardian*, September 10, 2014.

Apter, Michael J. *The Experience of Motivation: The Theory of Psychological Reversals*. New York: Academic Press, 1982.

Arendt, H. *The Human Condition*. Chicago: University of Chicago Press, 1998.

Arnett, J. J. "The Psychology of Globalization." *American Psychologist* 57, no. 10 (2002): 774–783.

Aronson, D. "How Congress Devastated Congo." *New York Times*, August 7, 2011.

Arthurs, E. "Travelodge Research Shows Over a Third of British Adults Still Sleep with a Teddy Bear." CISION PR Newswire. https://www.prnewswire.com/news-releases

/travelodge-research-shows-over-a-third-of-british-adults-still-sleep-with-a-teddy-bear-139427998.html (accessed May 13, 2016).

Ash, J. *Phase Media: Space, Time and the Politics of Smart Objects*. London: Bloomsbury, 2017.

Ashby, M., and K. Johnson. *Materials and Design: The Art and Science of Material Selection in Product Design*. Oxford: Butterworth-Heinemann, 2002.

Bafilemba F., T. Mueller, and S. Lezhnev. "The Impact of Dodd-Frank and Conflict Minerals Reforms on Eastern Congo's Conflict." *Enough Report*, Washington, 2014.

Baker, L. "How Fashion Can Stop Ruining the Planet." BBC. https://www.bbc.com/culture/story/20180419-how-fashion-can-stop-ruining-the-planet (accessed April 20, 2018).

Baker-Brown, D. *The Re-use Atlas: A Designer's Guide towards a Circular Economy*. London: RIBA Publishing, 2017.

Bakker, C., M. C. den Hollander, E. van Hinte, and Y. Zijlstra. *Products That Last: Product Design for Circular Business Models*. TU Delft Library, 2014.

Ballester, M. "From Ownership to Service: A New Fairphone Pilot Just for Companies." Fairphone. https://www.fairphone.com/en/2018/01/08/from-ownership-to-service-new-fairphone-pilot-for-companies (accessed January 8, 2018).

Banks, J. A., and C. A. McGee. *Multicultural Education: Issues and Perspectives*. Needham Heights, MA: Allyn & Bacon, 1989.

Bateson, G. *Mind and Nature: A Necessary Unity*. New York: E. P. Dutton, 1979.

Bateson, G. *Steps to an Ecology of Mind: Collected Essays in Anthropology, Psychiatry, Evolution, and Epistemology*. Chicago: University of Chicago Press, 2000.

Baxter, W., and P. Childs. "Designing Circular Possessions." In *Routledge Handbook of Sustainable Product Design*, ed. J. Chapman, 391–404. Oxon: Routledge, 2017.

Bayley, S. "What Makes a Design Classic?" *Independent*, August 27, 1999.

Belk, R. "Possessions and the Extended Self." *Journal of Consumer Research* 15, no. 2 (1988): 139–168.

Belk, R. "Three Scales to Measure Constructs Related to Materialism: Reliability, Validity, and Relationships to Measures of Happiness." *Advances in Consumer Research* 11 (1984): 291–297.

Belk, R. "Why Not Share Rather than Own?" *Annals of the American Academy of Political and Social Science* 611, no. 1 (May 2007): 126–140.

Benjamin, R. *Captivating Technology: Race, Carceral Technoscience, and Liberatory Imagination in Everyday Life*. Durham, NC: Duke University Press, 2019.

Bibliography

Benson, S. "The Myth of Sustainable Fashion." *Huffington Post*, February 7, 2020.

Berners-Lee, M. *How Bad Are Bananas? The Carbon Footprint of Everything*. London: Profile Books, 2010.

Birney, A. "System Change Capabilities: What Are the Capabilities We Need for System Change?" School of System Change, Forum for the Future. https://www.forumforthefuture.org/system-change-capabilities (accessed November 23, 2018).

Bockock, R. *Consumption*. Oxon: Routledge, 1993.

Boehnert, J. *Design, Ecology, Politics: Towards the Ecocene*. London: Bloomsbury, 2018.

Botsman, R. *What's Mine Is Yours: The Rise of Collaborative Consumption*. New York: Collins, 2011.

Bourdieu, P. *Distinction: A Social Critique of the Judgement of Taste*. Oxon: Routledge, 1984.

Bourdieu, P. "The Forms of Capital." In *Handbook of Theory of Research for the Sociology of Education*, ed. J. G. Richardson. Westport, CT: Greenwood, 1986.

Bowker, G. C., and S. L. Star. *Sorting Things Out: Classification and Its Consequences*. Cambridge, MA: MIT Press, 2000.

Boydston, J. A. *The Later Works of John Dewey*. Vol. 1, *1925–1953: 1925, Experience and Nature*. Carbondale: Southern Illinois University Press, 2008.

Boyer, B., J. W. Cook, M. Steinberg, and Helsinki Design Lab. *In Studio: Recipes for Systemic Change*. Helsinki: Sitra, 2011.

Bruner, J. *Acts of Meaning: Four Lectures on Mind and Culture*. The Jerusalem-Harvard Lectures, book 3. Cambridge, MA: Harvard University Press, 1990.

Campbell, E. "Design Classics: Unequivocal, Tangible, Iconic?" *RSA Design and Society Blog*. https://www.designandsociety.wordpress.com/2009/01/20/design-classics-unequivocal-tangible-iconic (accessed April 15, 2015).

Cannon, S. *Popular Beliefs and Superstitions from Utah*. Salt Lake City: University of Utah Press, 1984.

Chapman, J. *Emotionally Durable Design: Objects, Experiences and Empathy*. Oxon: Routledge, 2015.

Chapman, J. "Hadal or Epipelagic? The Depths, and Shallows, of Material Experience." In *Product Lifetimes and the Environment Conference Proceedings, 17–19 June, 2015—Nottingham, UK*, ed. T. Cooper, N. Braithwaite, M. Moreno, and G. Salvia, 57–61. Nottingham: Nottingham Trent University, CADBE, 2015.

Chapman, J., and G. Marmont. "The Temporal Fallacy: Design and Emotional Obsolescence." In *Routledge Handbook of Sustainable Design*, ed. R. B. Egenhoefer. Oxon: Routledge, 2018.

Chin, E. *My Life with Things: The Consumer Diaries*. Durham, NC: Duke University Press, 2016.

Cialdini, R. D. "Crafting Normative Messages to Protect the Environment." *Current Directions in Psychological Science* 12, no. 4 (2003): 105–109.

Clandinin, D. J., and F. M. Connelly. *Narrative Inquiry: Experience and Story in Qualitative Research*. San Francisco: Jossey-Bass, 2000.

Conway, R., I. Masters, and J. Thorold. *From Design Thinking to Systems Change: How to Invest in Innovation for Social Impact*. London: RSA Action and Research Centre, 2007.

Cooper, T. "Which Way to Turn? Product Longevity and Business Dilemmas in the Circular Economy." In *Routledge Handbook of Sustainable Product Design*, ed. J. Chapman, 405–422. Oxon: Routledge, 2017.

Crawford, K., and V. Joler. "Anatomy of an AI System: The Amazon Echo as an Anatomical Map of Human Labor, Data and Planetary Resources." New York: Share Lab/AI Now Institute, 2018.

Crewe, L. *The Geographies of Fashion: Consumption, Space, and Value*. London: Bloomsbury, 2017.

Csikszentmihalyi, M., and E. Rochberg-Halton. *The Meaning of Things: Domestic Symbols and the Self*. Cambridge: Cambridge University Press, 1981.

Cupchick, G. "The Half-Life of a Sustainable Emotion: Searching for Meaning in Product Usage." In *Routledge Handbook of Sustainable Product Design*, ed. J. Chapman, 25–40. Oxon: Routledge, 2017.

Cuthbertson, A. "I'm Not That Superstitious." Studying Religion in Culture. https://www.religion.ua.edu/blog/2016/01/27/im-not-that-superstitious (accessed March 12, 2017).

Dalhammar C., and J. Luth Richter. "Options for Lifetime Labeling: Design, Scope and Consumer Interfaces." In *Product Lifetimes and the Environment Conference Proceedings, 17–19 June, 2015—Nottingham, UK*, ed. T. Cooper, N. Braithwaite, M. Moreno, and G. Salvia, 461–463. Nottingham: Nottingham Trent University, CADBE, 2015.

Dall, A., and G. Smith. "The Sympathy of Things." BBC Sounds. https://www.bbc.co.uk/sounds/play/m0001188 (accessed April 21, 2019).

Darwin, C. *On the Origin of Species: By Means of Natural Selection*. London: Dover, 2006.

Davies, K., and L. Young. *Tales from the Dark Side of the City: The Breastmilk of the Volcano Bolivia and the Atacama Desert Expedition*. London: Unknown Fields, 2016.

Desmet, P. "Designing Emotions." PhD thesis, Faculty of Industrial Design Engineering, TU Delft, the Netherlands, 2002.

Desmet, P., and P. Hekkert. "Framework of Product Experience." *International Journal of Design* 1, no. 1 (2007): 57–66.

Dewey, J. *Theory of Valuation*. Chicago: University of Chicago Press, 1939.

Diderot, D. "Regrets for My Old Dressing Gown, or A Warning to Those Who Have More Taste than Fortune." Translated by M. Abidor for marxists.org (2005). In *Oeuvres complètes*, vol. 4. Paris: Garnier Frères, 1875.

DiGangi, J., and J. Strakova. "Toxic Toy or Toxic Waste: Recycling Pops into New Products." IPEN. https://ipen.org/sites/default/files/documents/toxic_toy_or_toxic_waste_2015_10-en.pdf (accessed October 31, 2017).

Dittmar, H., and L. Pepper. "To Have Is to Be: Materialism and Person Perception in Working-Class and Middle-Class British Adolescents." *Journal of Economic Psychology* 15, no. 2 (1994): 233–251.

Dostoyevsky, F. *The House of the Dead*. Mineola, NY: Dover, 2004.

Douglas, M., and B. Isherwood. *The World of Goods: Towards an Anthropology of Consumption*. London: Allen Lane, 1979.

Downe, L. "Good Services Are Verbs, Bad Services Are Nouns." Design in Government. https://www.designnotes.blog.gov.uk/2015/06/22/good-services-are-verbs-2 (accessed July 4, 2017).

Dubberly, H., U. Haque, and P. Pangaro. "What Is Interaction? Are There Different Types?" Dubberly Design Office. http://www.dubberly.com/articles/what-is-interaction.html (accessed January 7, 2019).

Dubberly, H., and P. Pangaro. "Cybernetics and Design: Conversations for Action." Dubberly Design Office. https://www.dubberly.com/articles/cybernetics-and-design.html (accessed November 1, 2015).

Dunne, A., and F. Raby. *Design Noir: The Secret Life of Electronic Objects*. London: Birkhauser, 2001.

Dunne, A., and F. Raby. *Speculative Everything: Design, Fiction, and Social Dreaming*. Cambridge, MA: MIT Press, 2013.

Dykstra, J. A. "Why Millennials Don't Want to Buy Stuff." Fast Company. https://www.fastcompany.com/1842581/why-millennials-dont-want-buy-stuff (accessed July 30, 2014).

Ehrenreich, B. *Bright-Sided: How Positive Thinking Is Undermining America*. London: Picador, 2010.

Ellen MacArthur Foundation. "Towards the Circular Economy: Economic and Business Rationale for an Accelerated Transition." Ellen MacArthur Foundation. https://www.ellenmacarthurfoundation.org/assets/downloads/publications/Ellen-MacArthur-Foundation-Towards-the-Circular-Economy-vol.1.pdf (accessed July 23, 2015).

Ellis, E. "Anthropocene." The Encyclopedia of Earth. https://editors.eol.org/eoearth/wiki/Anthropocene (accessed May 5, 2019).

Escobar, A. *Designs for the Pluriverse: Radical Interdependence, Autonomy, and the Making of Worlds*. Durham, NC: Duke University Press, 2018.

European Commission. "Closing the Loop: An EU Action Plan for the Circular Economy." *Communication from the Commission to the European Parliament, the Council, the European Economic and Social Committee and the Committee of the Regions*. Brussels, 2015.

Evans-Pritchard, E. E. *Witchcraft, Magic, and Oracles among the Azande*. Oxford: Clarendon Press, 1937.

Federal Trade Commission. "Internet of Things: Privacy and Security in a Connected World." FTC Staff Report. https://www.ftc.gov/system/files/documents/reports/federal-trade-commission-staff-report-november-2013-workshop-entitled-internet-things-privacy/150127iotrpt.pdf (accessed August 9, 2019).

Ferrufino, A. G. "Building a Foundation for More Responsible Gold Mining in Uganda." Fairphone. https://www.fairphone.com/en/2018/09/11/building-foundation-for-responsible-gold-mining-in-uganda (accessed September 18, 2018).

Festinger, L. *A Theory of Cognitive Dissonance*. Stanford, CA: Stanford University Press, 1957.

Fitzpatrick, C. "Conflict Minerals and the Politics of Stuff." In *The Routledge Handbook of Sustainable Product Design*, ed. J. Chapman, 197–205. Oxon: Routledge, 2017.

Fokkinga, S. "Design –|+ Negative Emotions for Positive Experiences." PhD diss., TU Delft, the Netherlands, 2015.

Fokkinga, S., and P. Desmet. "Darker Shades of Joy: The Role of Negative Emotion in Rich Product Experiences." *Design Issues* 28, no. 4 (October 2012): 42–56.

Fokkinga, S., P. Desmet, and J. Hoonhout. "The Dark Side of Enjoyment: Using Negative Emotions to Design for Rich User Experiences." In *Proceedings—Design & Emotion 2010*, Chicago, 2010. https://www.zenodo.org/record/2596843#.XUrMAJNKjdd (accessed May 5, 2019).

Forrester, J. W. "Counterintuitive Behavior of Social Systems." *Theory and Decision* 2, no. 2 (January 1971): 109–140.

Foster, K. R., and H. Kokko. "The Evolution of Superstitious and Superstition-like Behaviour." *Biological Sciences* 276, no. 1654 (2008): 31–37.

Foster, N. "The Future of Things." Hello Fosta. https://www.hellofosta.com/writing/the-future-of-things (accessed July 28, 2019).

Frijda, N. H. *The Emotions: Studies in Emotion and Social Interaction.* Cambridge: Cambridge University Press, 1987.

Fry, T. *New Design Philosophy: An Introduction to Defuturing.* Sydney: University of New South Wales Press, 1999.

Gapper, J. "Facebook Faces the Tragedy of the Commons." *Financial Times*, November 29, 2017.

Gartner, K. "Consumerism, Mass Extinction and Our Throw-Away Society." The Art Of. https://www.theartof.com/articles/consumerism-mass-extinction-and-our-throw-away-society (accessed October 24, 2016).

Gauntlett, D. *Making Is Connecting.* Cambridge: Polity, 2011.

Georges, R. A., and M. O. Jones. *Folkloristics: An Introduction.* Bloomington: Indiana University Press, 1995.

Giaccardi, E., and E. Karana. "Foundations of Materials Experience: An Approach for HCI." In *Proceedings of the 33rd SIGCHI Conference on Human Factors in Computing Systems*, 2447–2456. New York: ACM, 2015.

Giaccardi, E., E. Karana, H. Robbins, and P. D'Olivo. "Growing Traces on Objects of Daily Use: A Product Design Perspective for HCI." *DIS '14: Proceedings of the 2014 Conference on Designing Interactive Systems*, Vancouver, June 21–25, 2014, 473–482.

Gibbs, R. W. "Understanding and Literal Meaning." *Cognitive Science* 13 (1989): 243–251.

Gleick, P. "Water and Conflict." *International Security* 18, no. 1 (Summer 1993): 79–112.

Global Fashion Agenda. "Pulse of the Fashion Industry 2019." GFA. https://globalfashionagenda.com/initiatives/pulse/# (accessed May 29, 2019).

Grant, K. "Vivienne Westwood: Everyone Buys Too Many Clothes." *Daily Telegraph*, September 16, 2013.

Gregson, N., A. Metcalfe, and L. Crewe. "Practices of Object Maintenance and Repair: How Consumers Attend to Consumer Objects within the Home." *Journal of Consumer Culture* 9, no. 2 (June 2009): 248–272.

Grendler, P. F. "The Enigma of 'Wisdom' in Pierre Charron." *Romance Notes* 4, no. 1 (Autumn 1962): 46–50.

Griggs Lawrence, R. "Wabi-Sabi: The Art of Imperfection." *Utne Reader*, September–October 2001. https://www.utne.com/mind-and-body/wabi-sabi (accessed December 29, 2016).

Grohol, J. M. "Do You Still Have a Security Blanket?" PsychCentral. https://www.psychcentral.com/blog/do-you-still-have-a-security-blanket (accessed October 20, 2018).

Guber, M., M. McDonough, M. Kausch, S. Glass, and A. Gullingsrud. "Safe and Circular by Design: Making Positive Material Choices." ThinkDIF. https://www.thinkdif.co/sessions/safe-circular-by-design-making-postive-material-choices (accessed January 7, 2019).

Hahn, H. P., K. Hayes, and A. Kacapor. "Breaking the Chain: Ending the Supply of Child-Mined Minerals." Pact. https://www.pactworld.org/library/breaking-chain-ending-supply-child-mined-minerals (accessed October 27, 2018).

Haraway, D. *Staying with the Trouble: Making Kin in the Chthulucene*. Durham, NC: Duke University Press, 2016.

Hardin, G. "The Tragedy of the Commons." *Science* 162, no. 3859 (1968): 1243–1248.

Haskins, C. "AirPods Are a Tragedy." *Vice*, May 6, 2019. https://www.vice.com/en_us/article/neaz3d/airpods-are-a-tragedy (accessed July 2, 2019).

Hassenzahl, M. *Experience Design: Technology for All the Right Reasons*. Synthesis Lectures on Human-Centered Informatics. San Francisco: Morgan and Claypool, 2010.

Hassenzahl, M. "User Experience and Experience Design." In *The Encyclopedia of Human-Computer Interaction*, 2nd ed. Interaction Design Foundation. https://www.interaction-design.org/literature/book/the-encyclopedia-of-human-computer-interaction-2nd-ed (accessed July 10, 2019).

Hawthorne, N. *The Scarlet Letter*. With an introduction by George Parsons Lathrop. Scotts Valley, CA: CreateSpace Independent Publishing Platform, 2018.

Hayes, R. *Rethinking Society from the Ground Up*. London: Alliance for Sustainability and Prosperity, 2013.

Hebert, O. "Designing the Next Fairphone from the Inside Out." Fairphone. https://www.fairphone.com/en/2015/05/13/designing-the-next-fairphone-from-the-inside-out (accessed February 23, 2017).

Heidegger, M. *Being and Time*. New York: Harper Perennial Modern Classics, 2008.

Hekkert, P. "Design Aesthetics: Principles of Pleasure in Product Design." *Psychology Science* 48, no. 2 (2006): 157–172.

Her Majesty's Government. *Our Waste, Our Resources: A Strategy for England*. London: Defra, Resources and Waste Strategy Team, 2018.

Heskett, J. *Toothpicks and Logos: Design in Everyday Life*. Oxford: Oxford University Press, 2003.

Hill, D. *Dark Matter and Trojan Horses: A Strategic Design Vocabulary*. Helsinki: Strelka Press, 2014.

Hill, M. "How to Buy Less, Choose Well and Make It Last." Good on You. https://www.goodonyou.eco/how-to-buy-less-choose-well-and-make-it-last (accessed August 1, 2019).

Hirschman, E. C., and M. B. Holbrook. "Hedonic Consumption: Emerging Concepts, Methods and Propositions." *Journal of Marketing* 46, no. 3 (Summer 1982): 92–101.

Hood, B. *The Science of Superstition: How the Developing Brain Creates Supernatural Beliefs*. New York: HarperOne, 2010.

Horton, R. "African Traditional Thought and Western Science: Part I. From Tradition to Science." *Africa: Journal of the International African Institute* 37, no. 1 (1967): 50–71.

Houston, L., S. J. Jackson, D. K. Rosner, S. I. Ahmed, M. Young, and L. Kang. "Values in Repair." In *CHI '16: Proceedings of the 2016 CHI Conference on Human Factors in Computing Systems*, San Jose, CA, May 7–12, 2016, 1403–1414. New York: Association for Computing Machinery, 2016.

Hsu, F. L. K. "The Self in Cross-Cultural Perspective." In *Culture and Self: Asian and Western Perspectives*, ed. A. J. Marsella, G. deVos, and F. L. K. Hsu, 24–55. New York: Tavistock Publications, 1985.

Ingold, T. *Making*. Oxon: Routledge, 2013.

Ingold, T. "The Temporality of the Landscape." *World Archaeology* 25, no. 2 (1993): 152–174.

Ingold, T. "Towards an Ecology of Materials." *Annual Review of Anthropology* 41, no. 1 (2012): 427–442.

Irwin, T. "The Emerging Transition Design Approach." *Proceedings*, Design Research Society Conference 2018, Limerick, 25–28 June, 2018, 968–989.

Irwin, T. "Transition Design Preface." *Cuaderno 73, Transition Design Monograph* 19 (July 2018): 19–26.

Jarrett, C. "Why Are We So Attached to Our Things?" TED Talk. https://www.scienceandnonduality.com/why-are-we-so-attached-to-our-things (accessed January 30, 2017).

Juniper, A. *Wabi-Sabi: The Japanese Art of Impermanence*. North Clarendon, VT: Tuttle Publishing, 2003.

Kalantidou, E. "Handled with Care: Repair and Share as Waste Management Strategies and Community Sustaining Practices." In *Product Lifetimes and the Environment Conference Proceedings, 17–19 June, 2015—Nottingham, UK*, ed. T. Cooper, N. Braithwaite, M. Moreno, and G. Salvia, 158–165. Nottingham: Nottingham Trent University, CADBE, 2015.

Karana, E. "How Do Materials Obtain Their Meanings?" *Journal of the Faculty of Architecture Middle East Technical University* 27, no. 2 (2010): 271–285.

Karana, E., E. Giaccardi, and V. Rognoli. "Materially Yours." In *Routledge Handbook of Sustainable Product Design*, ed. J. Chapman, 206–221. Oxon: Routledge, 2017.

Kasser, T., R. M. Ryan, C. E. Couchman, and K. M. Sheldon. "Materialistic Values: Their Causes and Consequences." In *Psychology and Consumer Culture: The Struggle for a Good Life in a Materialistic World*, ed. T. Kasser and A. D. Kanner, 11–28. Washington, DC: American Psychological Association, 2004.

Kelemen, D. "The Scope of Teleological Thinking in Preschool Children." *Cognition* 70 (1999): 241–272.

Keulemans, G., N. Rubenis, and A. Marks. "Object Therapy: Critical Design and Methodologies of Human Research in Transformative Repair." *PLATE Conference Proceedings*, Delft, November 8–10, 2017, 186–191.

Kilborn, P. T. "Splurge: Why Americans Can't Stop Buying Stuff." *New York Times*, June 21, 1998.

Kluckhohn, C., and W. H. Kelly. "The Concept of Culture." In *The Science of Man in the World Culture*, ed. R. Linton, 78–105. New York, 1945.

Kok, L., G. Wurpel, and A. Ten Wolde. *Unleashing the Power of the Circular Economy*. Amsterdam: IMSA, 2013.

Koren, L. *Wabi-Sabi for Artists, Designers, Poets and Philosophers*. Point Reyes, CA: Imperfect Publishing, 2008.

Korzybski, A. *Science and Sanity: A Non-Aristotelian System and Its Necessity for Rigour in Mathematics and Physics*. Oxford: APA, 1933.

Kossoff, G. "Cosmopolitan Localism: The Planetary Networking of Everyday Life in Place." *Cuaderno 73, Transition Design Monograph* 19 (July 2018): 51–66.

Krippendorff, K., and R. Butter. "Semantics: Meanings and Contexts of Artifacts." In *Product Experience*, ed. H. Schifferstein and P. Hekkert, 353–375. Amsterdam: Elsevier, 2008.

Kukla, A. "Antirealist Explanations of the Success of Science." *Philosophy of Science* 63, no. 1 (1996): 298–305.

Leach, B., and S. Yanagi. *The Unknown Craftsman: A Japanese Insight into Beauty.* Tokyo: Kodansha International, 2013.

Le Billion, P. "Getting It Done: Instruments of Enforcement." In *Natural Resources and Violent Conflict*, ed. I. Bannon and P. Collier, 215–286. Washington, DC: World Bank, 2013.

Leontiev, D. *Positive Psychology in Search for Meaning.* Oxon: Routledge, 2014.

Lewis, G. "The Look of Magic." *Man: The Journal of the Royal Anthropological Institute* 21, no. 3 (September 1986): 414–437.

Li, S. "Skipping the 13th Floor." *Atlantic*, February 13, 2015. https://www.theatlantic.com/technology/archive/2015/02/skipping-the-13th-floor/385448 (accessed March 22, 2016).

Light, A., and C. Miskelly. "Sharing Economy vs. Sharing Cultures? Designing for Social, Economic and Environmental Good." *Interaction Design and Architecture(s)* 24 (2015): 49–62.

Lilley, D., G. Smalley, B. N. Bridgens, and G. T. Wilson. "Cosmetic Obsolescence? User Perceptions of New and Artificially Aged Materials." *Materials and Design* 101 (April 2016): 355–365.

Lloyd, W. F. *Two Lectures on the Checks to Population.* Oxford: Oxford University, 1833.

Lockton, D., and S. Candy. "A Vocabulary for Visions in Designing for Transition." *Cuadernos del Centro de Estudios de Diseño y Comunicación* 73 (2019): 27–49.

Loewy, R. "Personal Letter." *Times* (London), November 19, 1945.

Lyubomirsky, S. "Hedonic Adaptation to Positive and Negative Experiences." In *The Oxford Handbook of Stress, Health, and Coping*, ed. C. S. Carver, 233–239. Oxford: Oxford University Press, 2010.

Macfarlane, R. *Underland: A Deep Time Journey.* New York: W. W. Norton, 2019.

MacKay, D. J. C. *Sustainable Energy without the Hot Air.* Cambridge: UIT Cambridge Ltd., 2009.

Macleod, J. *Ends: Why We Overlook Endings for Humans, Products, Services and Digital; And Why We Shouldn't.* London: Joe Macleod, 2017.

Macpherson, C. B. *The Political Theory of Possessive Individualism: From Hobbes to Locke.* Reprint ed. Oxford: Oxford University Press, 1962.

Macvean, M. "For Many People, Gathering Possessions Is Just the Stuff of Life." *Los Angeles Times*, March 21, 2014.

Manzini, E., and J. Cullars. "Prometheus of the Everyday: The Ecology of the Artificial and the Designer's Responsibility." *Design Issues* 9, no. 1 (Autumn 1992): 5–20.

Matchar, E. "The Fight for the 'Right to Repair.'" *Smithsonian*. https://www.smithsonianmag.com/innovation/fight-right-repair-180959764/#f7D6EbUdOyRPt6VX.99 (accessed July 13, 2016).

Matthews, J. A., and H. Tan. "Circular Economy: Lessons from China." *Nature: International Weekly Journal of Science*, March 23, 2016. https://www.nature.com/news/circular-economy-lessons-from-china-1.19593 (accessed March 28, 2018).

Maynard Keynes, J. *The General Theory of Employment, Interest, and Money*. Seattle: Stellar Classics, 2016.

Mazé, R. "Politics of Designing Visions of the Future." *Journal of Future Studies* 23, no. 3 (March 2019): 23–38.

McCarthy, J., and P. Wright. "Technology as Experience." *Interactions* 11, no. 5 (October 2004): 42–43.

McCracken, G. "Culture and Consumption: A Theoretical Account of the Structure and Movement of the Cultural Meaning of Consumer Goods." *Journal of Consumer Research* 13, no. 1 (1986): 71–84.

McGilchrist, I. *The Master and His Emissary: The Divided Brain and the Making of the Western World*. New Haven, CT: Yale University Press, 2009.

Meadows, D. H. *Thinking in Systems: A Primer*. London: Earthscan, 2010.

Medina, M. *The World's Scavengers: Salvaging for Sustainable Consumption and Production*. New York: Altamira Press, 2007.

Merleau-Ponty, M. *Phenomenology of Perception*. Oxon: Routledge, 1962.

Michaelian, K. *Mental Time Travel: Episodic Memory and Our Knowledge of the Personal Past*. Cambridge, MA: MIT Press, 2016.

Michel, A. "Product Lifetimes through the Various Legal Approaches within the EU Context: Recent Initiatives against Planned Obsolescence." *PLATE Conference Proceedings*, Delft University of Technology, Delft, November 8–10, 2017, 266–269.

Miller, G. T. *Environmental Science: Sustaining the Earth*. London: Wadsworth Publishing, 1993.

Mishara, A. "Klaus Conrad (1905–1961): Delusional Mood, Psychosis and Beginning Schizophrenia." *Schizophrenia Bulletin* 36, no. 1 (2010): 9–13.

Misra, S., and J. Maxwell. "Three Keys to Unlocking Systems-Level Change." *Stanford Social Innovation Review*, April 29, 2016. https://www.ssir.org/articles/entry/three_keys_to_unlocking_systems_level_change (accessed June 3, 2019).

Moritz, M. "Open Property Regimes." *International Journal of the Commons* 10, no. 2 (2016): 688–708.

Morton, T. *Hyperobjects: Philosophy and Ecology after the End of the World.* Minneapolis: University of Minnesota Press, 2013.

Morton, T. "The Mesh." In *Environmental Criticism for the Twenty-First Century*, ed. S. LeMenager, T. Shewry, and K. Hiltner, 19–30. Oxon: Routledge, 2012.

My Modern Met. "Kintsugi: The Centuries-Old Art of Repairing Broken Pottery with Gold." *My Modern Met*. https://www.mymodernmet.com/kintsugi-kintsukuroi (accessed February 17, 2018).

Napier, A. D. *Making Things Better: A Workbook on Ritual, Cultural Values, and Environmental Behavior.* Oxford: Oxford University Press, 2013.

NASA. "Lucky Peanuts." NASA Solar System Exploration. https://solarsystem.nasa.gov/news/10022/lucky-peanuts (accessed June 28, 2019).

Nike. "Circularity: Guiding the Future of Design." Nike. https://www.nikecirculardesign.com (accessed August 1, 2019).

Norton, M., D. Mochon, and D. Ariely. "The IKEA Effect: When Labor Leads to Love." *Journal of Consumer Psychology* 22, no. 3 (September 2011): 453–460.

O'Brien, M. "Consumers, Waste and the 'Throwaway Society' Thesis: Some Observations on the Evidence." *International Journal of Applied Sociology* 3, no. 2 (2013): 19–27.

Odom, W., and J. Pierce. "Improving with Age: Designing Enduring Interactive Products." *CHI '09 Extended Abstracts*, CHI '09 Conference on Human Factors in Computing Systems, Boston, April 4–9, 2009, 3793–3798.

Oishi, E. "Semantic Meaning and Four Types of Speech Act." In *Perspectives on Dialogue in the New Millennium*, ed. P. Kühnlein, H. Rieser, and J. Benjamins. Amsterdam: John Benjamins, 2003.

Oster, M. *Reflections of the Spirit: Japanese Gardens in America.* London: Dutton Studio Books, 1994.

Overbeeke, K. C. J., and S. A. G. Wensveen. "From Perception to Experience, from Affordances to Irresistibles." In *Proceedings of 2003 International Conference on Designing Pleasurable Products and Interfaces*, ed. B. Hannington and J. Forlizzi, 92–97. Pittsburgh: ACM Press, 2003.

Owen, J. "'Possession Purgatory': Situated Divestment Delay in the Lifecycle of Objects." Paper presented at "Making and Mobilising Objects: People, Process and Place." University of Warwick, Interdisciplinary Postgraduate Conference, University of Warwick, Leicester, February 21, 2015.

Owen, J. "Waiting to Reach This Lull When the Sadness Has Become Slightly Less Desperate: Dealing with Material A/Effects: Distancing Memories and Emotion in Self-Storage." Paper presented at "The Senses and Spaces of Death, Dying and Remembering: Historical and Contemporary Perspectives" Conference, Leeds, March 27–28, 2018.

Palahniuk, C. *Fight Club*. New York: W. W. Norton, 2005.

Papanek, V. *Design for the Real World*. London: Thames & Hudson, 1972.

Payton, M. "Objects of Affection: And the Students Who Won't Leave Home without Them." *FDUMagazine*, Winter–Spring 2009. http://www.fdu.edu/newspubs/magazine/09ws/objects.html (accessed December 28, 2018).

Pedgley, O., B. Şener, D. Lilley, and B. Bridgens. "Embracing Material Surface Imperfections in Product Design." *International Journal of Design* 12, no. 3 (2018): 21–23.

Pennock, S. F. "On the Meaning of Meaning: What Are We Really Looking For?" *Positive Psychology*, November 24, 2019. https://www.positivepsychologyprogram.com/meaning (accessed September 23, 2017).

Perella, M. "Durable Duds: The Disruptive Startups Looking to Wear Out Fast Fashion." Sustainable Brands, May 23, 2016. https://www.sustainablebrands.com/read/product-service-design-innovation/durable-duds-the-disruptive-startups-looking-to-wear-out-fast-fashion (accessed May 2, 2018).

Peters, F. E. *Greek Philosophical Terms: A Historical Lexicon*. New York: NYU Press, 1967.

Phillips, R. *Futurekind: Design by and for the People*. London: Thames & Hudson, 2019.

Pierce, J. L., T. Kostova, and K. T. Dirks. "Toward a Theory of Psychological Ownership in Organizations." *Academy of Management Review* 26, no. 2 (2001): 298–310.

Pilkington, E. "Robin Wright Targets Congo's 'Conflict Minerals' Violence with New Campaign." *Guardian*, May 17, 2016.

Pine, B. J. *Mass Customization: The New Frontier in Business Competition*. Boston: Harvard Business Review Press, 1992.

Powell, R. R. *Wabi Sabi Simple: Create Beauty. Value Imperfection. Live Deeply*. Adams Media, 2004.

Raghavan, S. "Obama's Conflict Minerals Law Has Destroyed Everything, Say Congo Miners." *Guardian*, December 2, 2014.

Rams, D. *Less but Better*. New York: Gestalten, 2014.

Ravasi, D., and V. Rindova. "Creating Symbolic Value: A Cultural Perspective on Production and Exchange." *SSRN* 111, no. 4 (May 2004).

Reid-Henry, S. "Arturo Escobar: A Post-development Thinker to Be Reckoned With." *Guardian*, November 5, 2012.

Restart Project. "First Protest for the Right to Repair in Brussels." Restart. https://www.therestartproject.org/restart-project/first-ever-protest (accessed December 2, 2018).

Restart Project. "Move Slow and Fix Things." Restart. https://www.therestartproject.org/about (accessed February 26, 2017).

Restart Project. "Restart Radio: Strong Links between Thrift and Innovation, Past and Present." Restart. https://therestartproject.org/podcast/thrift-innovation-past-present (accessed December 3, 2018).

Richins, M. L., and S. Dawson. "A Consumer Values Orientation for Materialism and Its Measurement: Scale Development and Validation." *Journal of Consumer Research* 19 (1992): 303–361.

Rindova, V. P. "Cultural Consumption and Value Creation in Consumer Goods Technology Industries." *Academy of Management Proceedings* 2007, no. 1 (2017): article 1.

Rinne, A. "The Dark Side of the Sharing Economy." *World Economic Forum*, January 16, 2018. https://www.weforum.org/agenda/2018/01/the-dark-side-of-the-sharing-economy (accessed September 15, 2018).

Rittel, H. W. J., and M. M. Webber. "Dilemmas in a General Theory of Planning." *Policy Sciences* 4, no. 2 (1973): 155–169.

Rognoli, V., and E. Karana. "Towards a New Materials Aesthetic Based on Imperfection and Graceful Ageing." In *Materials Experience: Fundamentals of Materials and Design*, ed. E. Karana, O. Pedgley, and V. Rognoli, 145–154. Oxford: Butterworth-Heinemann, 2014.

Rokeach, M. *The Nature of Human Values*. New York: Free Press, 1973.

Rose, A. "How to Build Something That Lasts 10,000 Years." BBC Future, June 11, 2019. https://www.bbc.com/future/story/20190611-how-to-build-something-that-lasts-10000-years (accessed June 11, 2019).

Rose, D. *Enchanted Objects: Design, Human Desire, and the Internet of Things*. New York: Scribner, 2014.

Rosenbloom, S. "But Will It Make You Happy?" *New York Times*, August 7, 2010.

Rosner, D. K., M. Ikemiya, D. Kim, and K. Koch. "Designing with Traces." In *CHI '13 Proceedings of the SIGCHI Conference on Human Factors in Computing Systems*, Paris, April 27–May 2, 2013, 1649–1658.

RSA: The Great Recovery. "Designing for a Circular Economy." RSA. https://www.greatrecovery.org.uk/resources/designing-for-a-circular-economy (accessed May 23, 2016).

Saarinen, E. *The Search for Form in Art and Architecture*. New York: Dover, 1985.

Sacks, D. "The Sharing Economy." Fast Company. https://www.fastcompany.com/1747551/sharing-economy (accessed October 7, 2013).

Sarner, M. "The Age of Envy: How to Be Happy When Everyone Else's Life Looks Perfect." *Guardian*, October 9, 2018.

Sassen, S. "Who Owns Our Cities—and Why This Urban Takeover Should Concern Us All." *Guardian*, November 24, 2015.

Satyro, W. C., J. B. Sacomano, J. C. Contador, and R. Telles. "Planned Obsolescence or Planned Resource Depletion? A Sustainable Approach." *Journal of Cleaner Production* 195 (September 10, 2018): 744–752.

Sayers, D. *Creed or Chaos? and Other Essays in Popular Theology*. London: Religious Book Club, 1948.

Scherer, K. R., A. Shorr, and T. Johnstone, eds. *Appraisal Processes in Emotion: Theory, Methods, Research*. Oxford: Oxford University Press, 2001.

Schippers, M., and P. A. M. van Lange. "The Psychological Benefits of Superstitious Rituals in Top Sport." SSRN, ERIM Report Series Reference no. ERS-2005-071-ORG, December 2005.

Schor, J. B. *The Overspent American: Why We Want What We Don't Need*. New York: Harper Perennial, 1999.

Scott, B. *The Heretic's Guide to Global Finance: Hacking the Future of Money*. London: Pluto Press, 2013.

Senge, P. M. *The Fifth Discipline: The Art and Practice of a Learning Organisation*. London: Random House, 1990.

Shayler, M. "Deconstruction #2: Mobile Phone." In *The Great Recovery Project Report*. London: RSA, 2013.

Sheldon, K. M., and S. Lyubomirsky. "The Challenge of Staying Happier: Testing the Hedonic Adaptation Prevention Model." *Personality and Social Psychology Bulletin* 36, no. 5 (February 23, 2012): 670–680.

Sherif, M. *The Psychology of Social Norms*. New York: Harper, 1936.

Sherif, Y. S., and E. L. Rice. "The Search for Quality: The Case of Planned Obsolescence." *Microelectronics Reliability* 26, no. 1 (1986): 75–85.

Sieck, W. "Cultural Norms: Do They Matter?" Global Cognition. https://www.globalcognition.org/cultural-norms (accessed May 7, 2019).

Smagorinsky, P. "If Meaning Is Constructed, What's It Made From? Toward a Cultural Theory of Reading." *Review of Educational Research* 71, no. 1 (Spring 2001): 133–169.

Small, R. "Being, Becoming, and Time in Nietzsche." In *The Oxford Handbook of Nietzsche*, ed. J. Richardson and K. Gemes, 122–134. Oxford: Oxford University Press, 2013.

Smithers, R. "Fast Fashion: Britons to Buy 50m 'Throwaway Outfits' This Summer." *Guardian*, July 11, 2019.

Snare, F. "The Concept of Property." *American Philosophical Quarterly* 9, no. 2 (1972): 200–206.

SPACE10. "Closing the Loop: Welcome to the Circular Economy." Chap. 6 of *Imagine*. Medium, July 25, 2017. https://www.medium.com/space10-imagine/chapter-6-closing-the-loop-welcome-to-the-circular-economy-92665c9678d9 (accessed August 20, 2017).

SPACE10. "Putting the 'Fab' in Fabrication: Manufacturing in the Digital Age." Chap. 1 of *Imagine*, June 15, 2017. https://www.medium.com/space10-imagine/chapter-1-putting-the-fab-in-fabrication-manufacturing-in-the-digital-age-fc9c7670dc5c (accessed August 20, 2017).

St. George, J. "The Things They Carry: A Study of Transitional Object Use among U.S. Military Personnel during and after Deployment." Theses, Dissertations, and Projects, no. 973, Smith College. https://scholarworks.smith.edu/theses/973 (accessed January 3, 2018).

Stahel, W. R. "Durability, Function and Performance." In *Longer Lasting Products: Alternatives to the Throwaway Society*, ed. T. Cooper, 157–177. Farnham: Gower, 2010.

Stahel, W. R., and T. Jackson. "Durability and Optimal Utilisation: Product-Life Extension in the Service Economy." In *Clean Production Strategies*, ed. T. Jackson, 261–294. Boca Raton, FL: Lewis, 1993.

Stead, M., P. Coulton, and J. Lindley. "Spimes Not Things: A Design Manifesto for a Sustainable Internet of Things." EAD 2019, Proceedings, 2133–2152.

Stephens, D. "50 Things Every Man Should Own to Win at Life." TAM. https://www.theadultman.com/live-and-learn/things-every-man-should-own (accessed July 16, 2019).

Stewart, I., and J. Cohen. *Figments of Reality: The Evolution of the Curious Mind*. Cambridge: Cambridge University Press, 1997.

Suh, E., E. Diener, and F. Fujita. "Events and Subjective Well-Being: Only Recent Events Matter." *Journal of Personality and Social Psychology* 70, no. 5 (1996): 1091–1102.

Sullivan, M. "'Right to Repair' Legislation Has Now Been Introduced in 17 States." Fast Company. https://www.fastcompany.com/40518779/right-to-repair-legislation-has-now-been-introduced-in-17-states (accessed February 14, 2018).

Thomasson, E. "IKEA to Test Furniture Rental in 30 Countries." Reuters Sustainable Business. https://www.reuters.com/article/us-ikea-sustainability-idUSKCN1RF0WY (accessed April 7, 2019).

Thompson, C. "We Need a Fixer (Not Just a Maker) Movement." *Wired*, June 18, 2013. https://www.wired.com/2013/06/qq-thompson (accessed March 14, 2016).

Thompson, J. B. *Studies in the Theory of Ideology*. London: Polity, 1984.

Thwaites, T. *The Toaster Project, or A Heroic Attempt to Build a Simple Electric Appliance from Scratch*. Hudson, NY: Princeton Architectural Press, 2011.

Tonkinwise, C. "Beauty in Use." *Design Philosophy Papers* 1, no. 2 (2003): 73–82.

Troncoso, S. "Is Sharewashing the New Greenwashing?" *P2P Foundation Blog*. https://blog.p2pfoundation.net/is-sharewashing-the-new-greenwashing/2014/05/23 (Accessed July 16, 2017).

Tsai, M. "Making Mistakes." Graduate thesis. Carnegie Mellon University, School of Design, Pittsburgh, 2019.

Tsing, A. L. *The Mushroom at the End of the World: On the Possibility of Life in Capitalist Ruins*. Princeton, NJ: Princeton University Press, 2015.

Turkle, S. *Evocative Objects: Things We Think With*. Cambridge, MA: MIT Press, 2011.

Twemlow, A. *Sifting the Trash: A History of Design Criticism*. Cambridge, MA: MIT Press, 2017.

Unmade. "Our Vision." Unmade. https://www.unmade.com/vision/# (accessed August 5, 2019).

Valant, J. "Planned Obsolescence: Exploring the Issue." European Parliamentary Research Service. https://www.europarl.europa.eu/RegData/etudes/BRIE/2016/581999/EPRS_BRI(2016)581999_EN.pdf (accessed July 27, 2017).

Valdesolo, P. "Why 'Magical Thinking' Works for Some People." *Scientific American*, October 19, 2010. https://www.scientificamerican.com/article/superstitions-can-make-you (accessed January 6, 2018).

van Dyne, L., and J. L. Pierce. "Psychological Ownership and Feelings of Possession: Three Field Studies Predicting Employee Attitudes and Organizational Citizenship Behaviour." *Journal of Organizational Behavior* 25, no. 4 (2004): 439–459.

van Hinte, E., ed. *Eternally Yours: Visions on Product Endurance*. Rotterdam: 010 Publishers, 1997.

Vardouli, T. "Making Use: Attitudes to Human-Artifact Engagements." *Design Studies* 41, part A (November 2015): 137–161.

Vergou, A., M. Wong, and J. Morgan. "It's Not about Milk." Goldsmiths College, University of London. https://www.themilkhasturned.com/about (accessed July 2, 2019).

Vince, G. "The High Cost of Our Throwaway Culture." BBC Future, November 28, 2012. https://www.bbc.com/future/story/20121129-the-cost-of-our-throwaway-culture (accessed November 30, 2017).

Vyse, S. A. *Believing in Magic: The Psychology of Superstition*. Oxford: Oxford University Press, 2000.

Vyse, S. "Do Superstitious Rituals Work?" *Skeptical Inquirer* 42, no. 2 (2018): 32–34.

Wagner, S. "Haunted Possessions." Live About, August 20, 2018. https://www.liveabout.com/haunted-possessions-2596712 (accessed July 7, 2019).

Wahl, D. C. "A Brief History of Systems Science, Chaos and Complexity." Age of Awareness, July 18, 2019. https://www.medium.com/age-of-awareness/a-brief-history-of-systems-science-chaos-and-complexity-d9198b1a198d (accessed July 10, 2019).

Wallendorf, M., and E. J. Arnould. "My Favorite Things: A Cross-Cultural Inquiry into Object Attachment, Possessiveness, and Social Linkage." *Journal of Consumer Research* 14, no. 4 (March 1988): 531–547.

Waste and Resource Action Programme (WRAP). *Valuing Our Clothes: The Evidence Base*. Technical Report. Banbury: WRAP, 2012.

Waste and Resource Action Programme (WRAP). "WRAP Reveals the UK's £30 Billion Unused Wardrobe." WRAP, July 11, 2012. https://www.wrap.org.uk/content/wrap-reveals-uks-%C2%A330-billion-unused-wardrobe (accessed August 7, 2014).

Weiner, A. B. *Inalienable Possessions*. Berkeley: University of California Press, 1992.

White, K. Y. "Time to Throw Away Our Throwaway Culture." *Eco Business*, February 24, 2016.

Whiteley, N. "Toward a Throw-Away Culture: Consumerism, 'Style Obsolescence' and Cultural Theory in the 1950s and 1960s." *Oxford Art Journal* 10, no. 2 (1987): 3–27.

Wieser, H., N. Tröger, and R. Hübner. "The Consumers' Desired and Expected Product Lifetimes." In *Product Lifetimes and the Environment Conference Proceedings, 17–19 June, 2015—Nottingham, UK*, ed. T. Cooper, N. Braithwaite, M. Moreno, and G. Salvia. Nottingham: Nottingham Trent University, CADBE, 2015.

Williams, G. *Twenty One: 21 Designers for Twenty-First Century Britain*. London: V&A Publishing, 2012.

Williams, R. "The Analysis of Culture." In *Cultural Theory and Popular Culture: A Reader*, ed. J. Storey, 48–56. Athens: University of Georgia Press, 1998.

Wilson, T. D., and D. T. Gilbert. "Affective Forecasting: Knowing What to Want." *Current Directions in Psychological Science* 14, no. 3 (June 1, 2005): 131–134.

Winnicott, D. W. "Transitional Objects and Transitional Phenomena: A Study of the First Not-Me Possession." *International Journal of Psycho-analysis* 34, no. 2 (1953): 89–97.

Wittgenstein, L. *Philosophical Investigations*. Englewood Cliffs, NJ: Prentice Hall, 1999.

Wood, J. "The Culture of Academic Rigour: Does Design Research Really Need It?" *Design Journal* 3, no. 1 (2000): 44–57.

Wray, R. "In Just 25 Years, the Mobile Phone Has Transformed the Way We Communicate." *Guardian*, December 31, 2009.

Wright, P., J. Wallace, and J. McCarthy. "Aesthetics and Experience-Centered Design." *ACM Transactions on Computer-Human Interaction* 15, no. 4 (November 2008): article 18.

Zhang, M. "This Leica M-P 'Correspondent' Edition Was Designed by Lenny Kravitz." PetaPixel. https://www.petapixel.com/2015/02/24/this-leica-m-p-correspondent-edition-was-designed-by-lenny-kravitz (accessed August 11, 2016).

Ziemer, T. "Why Consumer Products Are Designed to Fail." All about Circuits, August 31, 2015. https://www.allaboutcircuits.com/news/why-consumer-products-are-designed-to-fail (accessed April 4, 2018).

Zientek, J. "Can Better Denim Change the World? Levi's Is Betting on It." Gear Patrol, August 8, 2019. https://www.gearpatrol.com/2019/08/08/paul-dillinger-levis-innovation/ (accessed August 10, 2019).

Zimmermann, K. A. "What Is Culture?" Live Science, July 13, 2017. https://www.livescience.com/21478-what-is-culture-definition-of-culture.html (accessed January 22, 2018).

Index

Acaroglu, Leyla, 160
Accra, Ghana, 54
Acts of Meaning (Bruner), 19
Adamson, Glenn, 162
Adaptation, hedonic, 6–12
Additive processes, in manufacturing, 143
Adidas, 65, 132
Adorno, Theodor, 25
Adventures in the Anthropocene (Vince), 15–16
Aesthetic level of product experience, 71, 72–73
Aesthetics of interaction, 71
Affective forecasting, 8
Affective level of experience, 67
Agamben, Giorgio, 114
Agency, and error, 92
Aging, of products, 97–101
 classics of design, 110–117
AI, and error, 91
AirBnB, 140
AirPods, 2
Aitchison, Jean, 19–20
Alessi, Alberto, 111
Amers, Morgan, 146
Andrew, Rachel, 5–6
Angola, 52
Anthropocene, 129, 130, 149
Apophenia, 79
Apple, 2, 41

Apple Watch, 16
Apter, Michael, 94–95
AR, 118
Arduino kits, 145
Arendt, Hannah, 99, 100
Ariely, Dan, 42
Arnould, Eric, 10, 36
Artificial Intelligence. *See* AI
Ash, James, 60–61
Assets, tangible and intangible, 138
Augmented reality (AR), 118
Azande people, 83

Bad houses, 86–87
Baker-Brown, Duncan, 130–131
Bakker, Conny, 154–155
Bateson, Gregory, 61–62
Baxter, Weston, 87, 138
Beliefs, supernatural, 78–84
Belk, Russell, 30
Belo Horizonte, Brazil, 54
Benjamin, Ruha, 163
Bialetti, Alfonso, 111
Birney, Anna, 160
Bocock, Robert, 6
Boehnert, Joanna, 131
Borrowing, 138–139
Botsman, Rachel, 125
Bourdieu, Pierre, 35
Boyer, Bryan, 155
Brazil, 54, 56

Bright-Sided (Ehrenreich), 92
Bromated fire retardants, 55
Bruner, Jerome, 19
Burns, Brian, 139

Cambodia, 52
Candy, Stuart, 61
Capital, forms of, 35
Cassiterite, 51–52
Charisma Machine, The (Ames), 146
Charron, Pierre, 93
Chile, 129
Chin, Elizabeth, 28–29
China, 56
Christopher Raeburn, 143
Circular design, 132
Circular economy, 58, 127, 129–133, 153–154, 156–157
Circular Economy Action Plan, 156–157
Classic long life approach to product life extension, 154–155
Classics of design, 110–117
Climate change, and material consumption, 7, 49
Clothing
 biodegradable, 65
 and possessive individualism, 125
 throw-away, 15
Code, digital, 119–120
Coltan, 51–52
Competency, display of, 43
Computer numerical control (CNC), 92, 143, 145
Concrete, 64–65
Confidence (protective) frame, 94
Conflict minerals, 50–54
Consumers and consumption, 151–152. *See also* Possessions; Products
 conspicuous, 7
 and deliberate curtailment of product life, 12–16
 glutony, 17
 hedonic adaptation, 6–12
 and norms, 4–5
 and planned obsolescence, 13–15
 and possessive individualism, 123–126
 and social media, 6
 and usership, 133–137
Contamination
 emotional, 87
 interaction, 138
Cook, Tim, 41
Cooper, Tim 152–153
Correspondent (Leica), 102–103
Cosmology, personal, 21
Coulton, Paul, 4
Crawford, Kate, 48–49, 137
Creed or Chaos (Sayers), 17
Crewe, Louise, 44, 77
Critical social theory, 25
Csikszentmihalyi, Mihaly, 37
Cultural capital (Bourdieu), 35
Cultural norms, 34–35
Culture, 32–33
Cupchik, Gerald, 8

Dall, Amica, 142–143
Damage, to products, 101–102
Dark objects, 84–87
Dash Marshall, 155
Data, digital, 119–120
Defuturing, 5
Democratic Republic of the Congo (DRC), 51, 53, 57
Den Hollander, Marcel, 154–155
Denmark, 83
Design (and designers), 157–161
 approaches to product life extension, 154–155
 circular, 132
 classics of, 110–117
 as a conversation about change and conservation, 61–62
 and First Law of Human Ecology, 110
 maladapted motivations, 1–2
 and meaning, 19–21

and product life, 150–151
repair as central to process, 40
sustainable design thinking, 17, 162–166
for uncertainty, 108
Design, Ecology, Politics (Boehnert), 131
Design for the Real World (Papanek), 1
Designs for the Pluriverse (Escobar), 4
Desire, upward creep of, 121
Desmet, Pieter, 71
Detatchment (protective) frame, 94
Developing and developed worlds, generalizations about, 31
Dewey, John, 66–67
Diderot, Denis, 120–121
Diderot effect, 120–121
Digital manufacturing, 142–143
Digital objects, 117–121
Digital systems, 136–137
Dillinger, Paul, 156
Dodd-Frank Act (2010), 52–53
Dostoyevsky, Fyodor, 8
Downe, Louise, 139
Dubberly, Hugh, 135
Durkheim, Émile, 23

Eames lounge chair, 115
Ecodesign Directive, 157–158
Economy
  circular, 58, 127, 129–133, 153–154
  linear, 130–131
  sharing, 137–141
Effort, 42–46
Ehrenreich, Barbara, 92
Electronics, consumer
  and conflict minerals, 50–54
  natural resources for, 47–50
Ellen MacArthur Foundation, 131
Emerging markets, 33
Emotional level of product experience, 71–72
Emotions, positive and negative, 89–93
Enchanted Objects (Rose), 73–74

Endings and endineers, 56
Ends (Macleod), 56
Endurance, of a product, 97–101
Engagements, 134
Envelopes of objects and interfaces, 60–61
Episodic memory, 7
Equipmentality, 24
Error, as a collaborator in making, 91–92
Escobar, Arturo, 4–5, 37–38
European Commission, Circular Economy Action Plan, 156–157
European Union, environmental policy focus, 155–156
E-waste, 54–59
Exchange principle, 25
Experience, 66–67, 88. *See also* Product experience
  and cultural meanings, 34
  and emotions, 94
  experiential levels, 67
  positive and negative, 89–90
  as process, 88–89
  rich, 87–95
Experience design, 69. *See also* Product experience

Fab City Global Initiative, 146
Facebook, 6, 145
Fairphone, 57–59
Festinger, Leon, 42
First Law of Human Ecology, 110
Fixer communities, 40–41
Fokkinga, Steven, 93, 94
Form, of hyperobjects, 108–109
Forrester, Jay Wright, 61
Foster, Nick, 118
Fragmentalism, 64
France, 157
Francis, Pope, 82
Free will, and error, 92
Fry, Tony, 4–5

Functional obsolescence, 13
Futurecraft Loop, 132
Futurekind: Design by and for the People (Phillips), 145–146

Gauntlett, 144–145
Germany, 115–116
Ghana, 54
Giaccardi, Elisa, 67
Gifts and gift giving, 25
  and effort, 42–43
Gillespie, Ed, 164
Gloucester (England) City Council, 86
Gold
  in consumer electronics, 50, 51–52
  Salmon Gold, 65
Google X, 118
Gregson, Nicky, 44
Gross domestic product (GDP), 17

Hadal zone, 75–76, 83
Happiness, and hedonic adaptation, 7
Haraway, Donna, 161
Hardin, Garrett, 110, 136
Haskins, Caroline, 2
Hassenzahl, Marc, 72–73, 74, 90
Hawthorne, Nathaniel, 112
Hayes, Randy, 63
Hazda people (Tanzania), 138
Hedonic adaptation, 6–12
Heidegger, Martin: influence in objected-oriented ontology, 108
Heidenmark, Pia, 153
Hekkert, Paul, 71–72
Heraclitus, 107–108
Hiut Denim Company, 132–133
Hoke, Jeff, 153
Hood, Bruce, 85–86
Horseradish, and meaning, 68
Human Ecology, First Law of, 110
Husserl, Edmund, 106
Hybrid approach to product life extension, 154–155

Hygenic contamination, 87
Hyperobjects, 108–109

Iceberg (metaphor), 59
Identity seeking, 7
  and stripping, 10
iFixit, 16
IKEA, 153
Immune system, psychological, 8
Inalienable Possessions (Weiner), 139
Indebtedness, and borrowing, 139
India, 56
Indifference, zone of, 114
Individualism, possessive, 123–126
Infrastructure, maintenance of, 39
Ingold, Tim, 131–132
Interaction, 135
  contamination, 138
Interpretive level of experience, 67
Intimacy, and things, 24–25
Instagram, 6, 145
Instrumental value, 23–24, 30
Intergovernmental Panel on Climate Change (IPCC), 49
Integrated circuit boards (ICBs), 52
Intel, 52
Internet of Things (IoT), 3–4
Intrinsic value, 23–24, 30
Irwin, Terry, 159
Ise, Japan, 57–58
Issigonis, Alex, 111–112
Italy, 115–116
Ivory Coast, 52

James, William, 31
Japan
  attitudes toward repair in, 44–45
  Wabi-Sabi aesthetic, 104–105
Jarrett, Christian, 138
Jeans, environmental impacts of a pair of, 59–60
Jet Propulsion Laboratory, 82
Joler, Vladan, 48–49, 136

Jōō, Takeeno, 104
Jordan, Michael, 84

Karana, Elvin, 67
Keret, Shira, 92
Keynes, John Maynard, 152
Kintsugi, 44
Korzybski, Alfred, 62
Kossoff, Gideon, 128
Kravitz Design, 102–103
Krippendorff, Klaus, 66

Languages, programming, 119–120
Laser cutters, 145
Lathrop, George Parsons, 112
Leather, mushroom, 65
Lebow, Victor, 46
Leica Camera AG, 102–103
Leontiev, Dmitry, 20
Levi Strauss & Co., 156
Lewis, Gilbert, 79
Light, Ann, 140–141
Lindley, Joseph, 4
Linear economy, 130–131
Lloyd, William Foster, 136
Lockton, Dan, 61
Loewy, Raymond, 13
London, Bernard, 13–14
Luck
  and dark objects, 84–85
  and superstition, 78–84
Lyubomirsky, Sonja, 7–8

McCartney, Stella, 65
McCracken, Grant, 120
Macfarlane, Robert, 47
McGilchrist, Iain, 73
Macleod, Joe, 56
Macpherson, C. B., 123–124
Magical thinking, 79
Maintenance, 38–42
  and effort, 42–47
  as progressive action, 44–45

Makers and making, 141–148
  and error, 91–92
Maker spaces, 145–146
Making Is Connecting (Gauntlett), 144–145
Making Things Better (Napier), 37
Manufacturing, digital, 142–143
Manzini, Ezio, 5
Marx, Karl, 23, 35
Mass customization, 143
Material incongruence, 65
Materials
  local or not local, 128–129
  and meaning, 64–68
Meadows, Donella, 159
Meaning, 19–21
  kinds of, 66
  and materials, 64–68
Meaning level of product experience, 71, 72
Memory, episodic, 73
Merleau-Ponty, Maurice, 106, 134
Mesh, 131–132
Metcalfe, Alan, 44
Methane, 49
Michaelian, Kourken, 73
Microsoft, 65
Minerals, conflict, 50–54
Mini (automobile), 111–112
Mining, 127
  of conflict minerals, 50–54
  urban, 127–128
Miskelly, Clodagh, 141
Mochon, Daniel, 42
Moka coffee maker, 111
Monolith (Keret), 92
Morton, Timothy, 108–109, 131
Motorola, 52
Museum of Modern Art (MoMA), 143
Mushroom at the End of the World: On the Possibility of Life in Capitalist Ruins, The (Tsing), 130

My Life with Things (Chin), 28–29
Mylo, 65

Napier, David, 37
Narrative inquiry, 26
National Aeronautics and Space
    Administration (NASA), 82
Natural resources
  and conflict minerals, 50–54
  and consumer electronics, 47–50
  and urban mining, 127–129
  and waste, 2
New Balance, 143–144
Nietzsche, Friedrich, 106
Nike, 144, 153
Nike by You, 144
Nilsen, Ingrid, 82
Norms
  cultural, 34–35
  and the uses of goods, 2–3
Norton, Michael, 42
No Wash Club, 132–133

Obama, Barack, 82
Objects. *See also* Possessions; Products
  contamination, 87
  and cultural formation, 33
  dark, 84–87
  digital, 117–121
  maintenance of, 38–42
  spatiotemporality of, 107–110
  and superstition, 78–84
  and subject/object dichotomy, 33–34, 43
  transitional, 80–81
Object handling sessions, 25–26
Objected-oriented ontology (OOO), 108–109
O'Brien, Martin, 18
Obsolescence
  planned, 13–15
  unplanned, 15

Opening Ceremony, 143
Optimism, overvalue of, 92–93
Overspent American, The (Schor), 7
Owen, Jen, 126
Ownership. *See* Possessions, personal

Packard, Vance, 13
Palahniuk, Chuck, 124
Papanek, Victor, 1–2
Pen, concrete, 65
Pendant light, concrete, 65
Performative level of experience, 67
*Phase Media: Space, Time and the Politics
    of Smart Objects* (Ash), 60–61
Phillips, Robert, 145–146
Placebo effect, and superstition, 79
Planned obsolescence,, 13–14
Pluriverse, 37–38
Political Theory of Possessive Individualism
    (Macpherson), 123–124
Possessions, personal. *See also* Objects;
    Products
  cherished, 23, 26–27
  and lasting sentiments about
    products, 114–115
  and lived experience, 26
  maintaining (repairing), 38–42
  and meaning, 19–21, 23–28
  and memory, 27–28
  ownership vs. usership, 139–140
  and possessive individualism, 123–126
  power of, 9–11
  products becoming, 23–31, 97
  and self, 29
  sharing, 137–141
  and stripping, 10
  and usership, 133–140
  and value creation, 23–24, 29–30, 150
Potatoes, native source of, 129
Pragmatic meaning, 20
Product experience, 69–74
  rich, 87–95

Production-oriented, fast-replacement system, 3–4
Products. *See also* Objections; Possessions
   aging, 101–107. 117–121
   classics, 110–117
   digital, 117–121
   endurance in, 97–101
   perfection and imperfection in, 102–104
   as spatiotemporally diffuse, 107–110
Protective frame, 94–95
Psychological obsolescence, 13
Pyschology, and hedonic adaptation, 7–8
Pyschology ownership theory, 30

Raeburn, Christopher, 143
Ranger 7, and peanuts, 82
Rams, Dieter, 162
Recycling, 56. *See also* Waste; Waste pickers and picking
Repair, 39–41, 44–45, 147–148
Repair cafés, 40–41
Replacement, as a norm, 3–4, 39
Restart Project, 147
Rethinking Society from the Ground Up (Hayes), 63
Rich experience, 87–95
Rikyū, Sen no, 104–105
Rissanen, Timo, 160
Rituals, and superstition, 79–80
Rochberg-Halton, Eugene, 37
Rose, David, 73–74
Royal Society of Chemistry (UK), 127
Royal Society Winton Prize for Science Books, 15
Russia, 83

Saarinen, Eliel, 109
Safety zone (protective) frame, 94

Said, Edward, 105
Salmon Gold, 65
Sayers, Dorothy L, 17
Scarlet Letter, The (Hawthorne), 112
Schor, Juliet, 7, 121
Schulz, Charles, 80
Screws, and assembly, 40
Self, and the meanings of things, 28–30
Selfridges, 143
Semantic meaning, 20
Senge, Peter, 159
Sensorial level of experience, 67
Serbia, 83
Sharing economy, 137–141
Shinto shrines (Ise, Japan), 57–58
Sierra Leone, 52
Smartphone, environmental impacts of production and use, 60, 62–63
Smith, Giles, 142–143
Solder, 52
Souvenirs, 29
SPACE10, 143
Spain, 83, 116–117
Spatiotemporalisty, 107–110
Stahel, Alter, 3
Standardization, and meanings across cultures, 37
Star, Susan Leigh, 3
Starck, Philipe, 115
Status, social, 35–36
Stead, Michael, 4
Streamlining, 13
Stripping, and identity, 10
Subtractive processes, in manufacturing, 143
Superstition, 78–84
   and dark object, 84–87
Supply chain optimization, 17
Sustainability, 5
   and design thinking, 17, 162–166
Swiss Army knife, 112

Sympathy of Things, The (radio program), 142

Talismans, 82–83
Tantalum, 51, 53, 57
Territorial contamination, 87
Theseus, 58
Thirteenth floor, 85
3D knitting, 144
3D printers and printing, 143, 145
3TG, 52–53, 65
Throughput, 17–18
   material, 19
Throwaway society, 31
Thwaites, Thomas, 146–147
Tiffany & Co., 65
Tin, 51, 53, 57
Toaster Project, The (Thwaites), 146–147
Toasters, 146–147
Tommy Hilfinger, 7
Tonkinwise, Cameron, 112–113
Transition Design, 159
Transitional objects, 80–81
Tremlow, Alice, 12–13
Tsai, Mary, 91
Tsing, Anna, 130
Tungsten, 52, 53
Turkle, Sherry, 27
22 Studio, 65
Twitter, and consumption, 6

Ubikubi, 65
Uganda, 57
United Kingdom, 115–116, 127, 129
United Nations (UN)
   Human Development Index, 51
   UNICEF, 51
United States, 115–116, 157
Universal statements, 31–32
Unmade, 143–144
Urban mines and mining, 127–133
Usership, 133–137

Utility, economic, 1
Utility contamination, 87

Value
   changes in, 150
   creation, 23–24
   orientation, 77
Van Abel, Bas, 58
Van Hinte, Ed, 154–155
Vice (magazine), 2
Victorinox, 112
Vince, Gaia, 15–16
Vyse, Stuart, 79–80

Waag Society, Open Design Lab, 58
Wabi-Sabi, 104–105
Wahl, Daniel Christian, 161
Wake-Up Light, 74
Wallendorf, Melanie, 10, 36
Walmart, 67
Waste, 151–152
   and consumer electronics, 49–50
   and design, 1–2, 17–18
   sociology of, 18–19
   "universality" of, 31–32
   and urban mines, 127
   waste pickers and picking, 54–59
Waste and Resources Action Programme (WRAP), 153
Waste Electrical and Electronic Equipment Directive (WEEE Directive), 156
Waste Makers, The (Packard), 13
Water jet cutting, 92, 143
"We," statements about, 31
Wear and tear, 99
Weber, Max, 23–24
Weiner, Annette, 139
West, Fred, 86
West, Rosemary, 86
West house, 86–87
Westwood, Vivienne, 162
Will, and endurance, 98

Williams, Raymond, 34
Winnicott, Donard, 80
Wittgenstein, Ludwig, 66
Wolframite, 51–52
Wood, Johm 99–100
Woods Hole Research Center,. 49

YouTube, 137

Zijlstra, Yvo, 154–155
Zone of indifference, 114